MBAA Practical Handbook for the Specialty Brewer

Volume 2

Fermentation, Cellaring, and Packaging Operations

Edited by Karl Ockert

BridgePort Brewing Company
Portland, Oregon

Master Brewers Association of the Americas

Cover photographs of brewery equipment (clockwise from upper left)
courtesy of Filtrox, Sierra Nevada Brewing Company, Christian Gresser
GmbH, and New Glarus Brewing Company (photograph by Cynthia Stalker)

Library of Congress Control Number: 2005931016
International Standard Book Numbers:
0-9770519-1-9 (v. 1)   *Raw Materials and Brewhouse Operations*
0-9770519-2-7 (v. 2)   *Fermentation, Cellaring, and Packaging Operations*
0-9770519-3-5 (v. 3)   *Brewing Engineering and Plant Operations*

Printed in the United States of America on acid-free paper

The Master Brewers Association of the Americas
3340 Pilot Knob Road
St. Paul, Minnesota 55121, U.S.A.

Contents

CHAPTER 5

Candace E. Wallin
Instructor, Master Brewers Program, University of California, Davis

Foreword

In 1944, four master brewers—Edward Vogel (Griesedieck Bros. Brewery, St. Louis, Missouri), Frank Schwaiger and Henry Leonhardt (Anheuser-Busch, St. Louis, Missouri), and J. Adolf Merten (Ems Brewing Company, East St. Louis, Illinois)—volunteered to write a manual for the brewery worker if the Master Brewers Association of the Americas would publish it. Two years later their labors produced *The Practical Brewer,* a work so fundamentally complete that it still has relevance sixty years later.

At a meeting of the Executive Committee of the Master Brewers Association of the Americas in January 2003 it was suggested that a new adaptation of *The Practical Brewer* be written in the original question-and-answer format. This endeavor would help the Association meet one of its purposes, to improve the art and science of brewing by disseminating information of value to its members, the profession, the brewing and associated industries, and the public.

The finished product should be useful for today's entire community of brewers, whether they work for craft or microbrewers or for large brewing companies or whether they are home brewers—all are certainly interested in expanding their knowledge of the art and science of brewing.

Jaime Jurado, of the Gambrinus Company, San Antonio, Texas, took up this suggestion and convinced Karl Ockert, of the BridgePort Brewing Company, Portland, Oregon, to be the editor-in-chief of the book. Karl recruited an excellent group of authors willing to volunteer their time and share their knowledge of brewing fundamentals with their brewing colleagues.

After much diligent work by everyone involved in this project, each chapter of this exciting new book, the *MBAA Practical Handbook for the Specialty Brewer,* has been written, edited, and reviewed and is ready to stand alongside *The Practical Brewer* as an exceptional resource of practical brewing fundamentals.

Like the authors of *The Practical Brewer,* published in 1946, the authors of the *MBAA Practical Handbook for the Specialty Brewer* are volunteers, who will receive no other reward than the thanks and appreciation of the Master Brewers Association of the Americas and the satisfaction of the completion of a job well done.

Frank J. Kirner

President, 2003
Master Brewers Association
of the Americas

Preface

The following pages were written by our colleagues in the spirit of camaraderie that sets the brewing industry apart from almost any other field. The authors wrote these chapters while continuing to meet the obligations of their busy professional schedules and their personal lives. None will collect a royalty or any other monetary payment for their efforts. They are all true professionals, and we benefit from the discussions they bring forth. That, to me, is what the MBAA is all about.

Each chapter is a distillation of the brewing knowledge that each writer possesses. The following chapters encompass not only the art and science of brewing as dogma but the education and experience that these writers have been exposed to during their careers.

It has been my honor and privilege to be a part of this project and to help the MBAA assemble this team of authors and then assist them in putting their works together for others to enjoy. In addition to the authors, I have had the pleasure of working with Frank Kirner, Inge Russell, Ray Klimovitz, Laura Harter, Gil Sanchez, Prof. Charlie Bamforth, Prof. Ludwig Narziss, and my friend and mentor, Prof. Emeritus Michael Lewis, all of whom have helped me with the challenging process of editing and refining this handbook.

I would like to thank the MBAA Districts Northwest, Texas, Eastern Canada, and New York, which helped sponsor the production of this book through their district treasuries.

Finally, I would especially like to thank Carlos Alvarez for allowing me to work on this book project and my wife, Carole Ockert, who allowed this project to take time away from our home life.

Karl Ockert

Editor-in-Chief

July 2005

Authors

Bob August is a graduate of California State University, Chico. He began his career in 1982 as a brewer with the Sierra Nevada Brewing Company and progressed to become director of packaging operations from 1984 to 2004. He now operates an independent consulting firm; Majestic Packaging Solutions, in Chico.

Stephen Bates started his career in 1967 as a packaging technician with the Blitz Weinhard Brewing Company, in Portland, Oregon, and in 1990 became assistant packaging manager. In 1999, he took the position of packaging manager for the BridgePort Brewing Company, in Portland, Oregon, in charge of one bottling line and one racking line. He was also the project manager for the development of a new bottle packaging facility at the Trumer Brauerei, in Berkeley, California.

Daniel Carey has a bachelor of science degree in food science and technology from the University of California, Davis, in 1982. He attended the Siebel Institute course in brewing technology in 1987 and was valedictorian, and he passed the Diploma Master Brewer Examination of the Institute of Brewing in 1992. He served his apprenticeship at the Ayinger Brewery, in Munich. He is presently the co-owner and brewmaster of the New Glarus Brewing Company, in New Glarus, Wisconsin. He previously designed, constructed, and operated numerous small breweries while working for JV NorthWest, and he was a production supervisor at Anheuser-Busch, in Ft. Collins, Colorado.

Ken Grossman founded the Sierra Nevada Brewing Company, in Chico, California, in 1980 and serves as its president. He studied chemistry and physics at Butte College and at California State University, Chico. He has been an active member of the Master Brewers Association of the Americas since the early 1980s and also serves on the technical committee of the Brewers Association of America.

Candace E. Wallin has more than 25 years of experience in the brewing industry, including 18 years at Miller Brewing Company, Milwaukee, where she worked as a microbiologist and then as a development brewer. She also gained four years of experience in the microbrewery segment of the industry as the microbiologist for Sudwerk Privatbrauerei Hübsch, in Davis, California. Currently, she is an instructor in brewing science and technology courses in the extension program of the University of California, Davis, and she manages the brewing laboratory and pilot brewery there. She is an associate member of the Institute of Brewing and Distilling and a member of the American Society of Brewing Chemists and the Master Brewers Association of the Americas.

Grant E. Wood is pilot plant manager for the Boston Beer Company. He received a B.S. in food science and technology from Texas A&M University in 1984 and started his brewing career with the Pearl Brewing Company and Lone Star Brewing Company, in San Antonio, Texas. He has been with the Boston Beer Company since 1995. A 1987 graduate of the Siebel Institute course in brewing technology, he has been involved in such diverse activities as malting barley and hops selection and product development. He has judged at the Great American Beer Festival, in Denver, Colorado, and the Brewing Industry International Awards, in Burton on Trent, England.

CHAPTER 1

Fermentation and Cellar Operations

Daniel Carey

New Glarus Brewing Company

Ken Grossman

Sierra Nevada Brewing Company

1. What is fermentation?

Alcoholic fermentation is the process of breaking down sugars into ethanol, carbon dioxide, and energy. The purpose of brewery fermentation is to produce a flavorful and moderately alcoholic beverage from relatively unpalatable wort. In the fermentation cellar, the brewer combines wort with brewer's yeast. Although brewer's yeast can respire, it prefers anaerobic conditions when growing in nutrient-rich wort. Some of the intermediate compounds and energy produced during fermentation are used for cell reproduction. The by-products of these processes are responsible for much of beer flavor.

Sound fermentation produces "green beer" that has the proper flavor and is free from microbiological contamination. It also yields sufficient viable yeast for the next fermentation cycle while leaving a sufficient cell count in the beer for proper conditioning during aging.

Brewing Yeast

2. What is brewer's yeast?

Brewer's yeast is a unicellular fungus that is generally classified into two broad families, commonly referred to as "ale yeast" for top-fermenting strains and "lager yeast" for bottom-fermenting strains. This distinction is made based on their typical flocculation characteristics, fermentation temperature profiles, and the types of beer being produced. These

1

distinctions are less relevant today because of changes in modern fermentation systems and the continual evolution of brewing methods.

3. What differentiates yeast strains?

There has been a significant amount of reclassification and debate in the identifying and naming of brewing strains. With the development of DNA technology and other advanced identification methods, additional changes will probably follow. To avoid confusion, we will continue to refer to these strains as they have been known to the brewing industry for years, although these names are now considered taxonomically incorrect. Most brewing strains are of the genus *Saccharomyces* and the species *cerevisiae* for ale yeast and *pastorianus* (formerly *carlsbergensis* or *uvarum*) for lager yeast. Some distinguishing characteristics include fermentation temperature and fermentation period. Lager yeast ferments at a lower temperature—typically in the 41–54°F (5–12°C) range and with a 5- to 12-day fermentation period (although some styles of beer are brewed using lager yeast at close-to-ale fermentation temperatures), has the ability to metabolize the sugar melibiose, and produces greater quantities of H_2S and other sulfur compounds. Ale yeast performs best at temperatures in the 59–72°F (15–22°C) range with a fairly rapid 3- to 5-day fermentation period and tends to produce more esters and higher alcohols.

4. Why are different strains used?

Hundreds of different yeast strains or variants, which were historically used or selected for their flavor contributions and fermentation characteristics, are utilized by breweries worldwide. Some brewers use a mixed culture of yeast strains, each of which may contribute a different fermentation attribute or production benefit. Most likely, these mixtures were in use before good differentiation techniques were developed. Even though the mixtures complicate yeast handling, brewers may be reluctant to change for fear of altering the character of the beer. Some styles of beer may include one strain for the main fermentation and a separate strain for bottle conditioning. Several specialty styles of beers are fermented with yeast strains that produce distinctive flavors and aromas, such as those used to produce some styles of wheat beer or the *Brettanomyces* strains used to ferment Belgian lambic beers. Some breweries produce a distinctive style of beer in which a combination of yeast and bacteria are used to achieve a unique flavor profile.

5. What is wild yeast?

Any strain other than the specific brewery culture yeast is typically referred to as wild yeast. Cross contamination of strains may occur in breweries utilizing multiple strains for different styles of beer. The nondesirable strain may be considered "wild," although most problems encountered with wild yeast result from nonbrewing strains. Many wild yeast strains are responsible for off-flavors, aromas (notably phenolic), hazes, and attenuation problems. Some wild yeast strains are referred to as "killer yeast" because they excrete toxins that inhibit the growth of culture yeast.

6. What are the sources of wild yeast contamination?

Wild yeast cells are typically airborne because they can attach to dust particles such as those from malt, dirt, or other agricultural products. During certain times of the year (such as when the weather is warm and fruit is ripening), the problem may be worse, although the cells can also be transported with raw materials, with food, or on bodies or clothing in any season. Good housekeeping and sanitation in and around the brewery are essential to minimize wild yeast growth and contamination.

7. How are wild strains identified?

Although some strains are very difficult to identify, various techniques have been developed (see Chapter 5, Volume 2). Many wild yeast strains can be detected by using media that contain a compound that inhibits the growth of brewing yeast and allows the wild yeast strain to multiply. Most wild strains can be detected by plating brink yeast or beer on three different media—one containing cupric sulfate, another containing fuchsin sulfate, and a third containing lysine. Identifying wild yeast from the same genus as the brewing strains requires other advanced methods and more extensive laboratory skills. Some wild strains that may not be considered true beer spoilers can still be problematic. Brewers who produce both lagers and ales can experience attenuation and flavor issues caused by cross contamination of the different brewing yeast strains. These strains can generally be differentiated by using media that change color on the basis of the ability of the lager strain to utilize melibiose, or the differences in cell morphology may be recognizable by microscopic examination of the yeasts.

8. What is the morphology and cytology of an individual yeast cell?

Brewer's yeast cells are generally round to oval or elliptical and range from 2.5 to 20 microns in diameter; the typical size is 5–10 microns. As the cells age, they tend to expand in size. Some strains have cells that are individual or in groups, and others tend to form clumps or chains. The main components of the yeast cell are the outer cell wall, the periplasm, and a permeable membrane, which contain the cytoplasm but allow the diffusion of nutrients and other necessary solutes. Inside the cell, the genetic material (DNA) is located primarily in the nucleus but is also contained in the mitochondria. The cell also houses one or more vacuoles containing a polyphosphate reserve. Also stored in the cytoplasm is the carbohydrate energy source, glycogen. When viewed in a microscope, the outer cell wall typically reveals both bud and birth scars.

9. How is microscopic examination used to evaluate a yeast sample?

Once a brewer is familiar with the appearance of a particular strain of yeast, a microscopic examination of the yeast culture may be useful in assessing the overall health of the population. Abnormal-looking or irregularly shaped cells can be an indication of cell stress, possibly indicating potential problems with wort composition, aeration, poor yeast handling, or fermentation conditions. Utilizing dyes or staining methods, the brewer can determine the overall viability of the culture visually in a basic bright-field microscope. A microscopic examination can sometimes be useful in uncovering wild yeast or bacterial infection, although if the contamination is readily observed in the microscope, it is probably too late to rectify the problem.

10. How do yeast cells multiply?

The yeast cell multiplies by cell division. A small bud forms on the "mother" and grows until its size matches that of the parent. When the bud is mature, it typically separates, forming an identical new cell or "daughter." Some strains do not readily separate and instead form chains or groups of cells. The rate of cell growth and division is dependent on many factors, including yeast strain, vitality, temperature, and the availability of nutrients and oxygen. In most brewery fermentations, a doubling of cells typically occurs in 2–6 hours. Healthy yeast shows heavy budding under the microscope during the first stages of fermentation.

11. How are yeast cultures obtained?

Large collections of brewing yeast cultures are maintained by numerous laboratories and universities throughout the world and are catalogued by various criteria such as fermentation rate, flocculation, and different flavor and aroma characteristics. Some of these databases are now available online, and searches can be made for a specific set of attributes. Many of these yeast strains have great pedigrees, having been proven performers in many breweries around the world. Most of these cultures can be purchased in several different forms. Most brewers propagate from a wort agar "slant," which can be stored under refrigeration for up to 6 months with little degradation. The working culture is propagated from the slant by utilizing a flamed loop to remove a small scraping of the deposited cell culture and inoculating a small amount of sterile wort (see the section on propagation later in this chapter). Working from a slant requires impeccable cleanliness and good technique to ensure that no contamination occurs. Some brewers choose to purchase an active liquid culture that minimizes some of the contamination risk inherent in starting from a slant. Liquid cultures must be immediately introduced into the propagation sequence, since once activated, the yeast cells have a limited life span in the slurry. Very small breweries can purchase quantities large enough to start a complete fermentation. Cultures can also be stored in liquid nitrogen for many years with little degradation. A limited number of brewing yeast strains have been made available commercially in a freeze-dried form. Although this technique has been accepted and works satisfactorily in other industries utilizing yeast (such as baking and winemaking), the forms marketed to small breweries and home brewers have had mixed success because of either contamination or performance problems.

Some multisite breweries and those engaged in contract production arrangements may propagate yeast at a central company brewing lab. The yeast can be distributed as a wet slurry on a regularly scheduled basis (sterile beer kegs have been used for transport), or the culture can be freeze-dried for ease of handling. Historically, some small breweries have relied on cooperative larger breweries to supply their pitching yeast in either wet slurry or pressed yeast cakes. With potential quality-control issues, consolidation in the industry, and competitive issues, this is probably not a viable strategy for brewers to rely on in the future.

12. How are yeast cultures isolated?

Several techniques for obtaining or isolating culture yeast involve selecting one or several cells from an existing fermentation and propagating them to a sufficient quantity to introduce back into the main plant. For example, this technique can be used to reselect for a particular variant, such as good flocculation, by selecting the culture yeast from a sample that exhibits that desired trait. Another method involves streaking out yeast from a typical fermentation on a petri dish and then isolating individual colonies based on morphological differences. Pilot fermentations must then be utilized to verify that flavor, analytical performance, or other desirable traits have been selected. These techniques must be carefully monitored to prevent selection of an unwanted variant. Since the character of the beer can be very dependent on yeast strain, it is probably a sound practice to have one or more offsite storage facilities maintain the culture in the event of an undesirable drift, widespread contamination, or other production catastrophe.

13. What is the yeast life cycle curve?

During fermentation, yeast cells follow a life cycle that can be divided into five phases: the lag phase, the growth phase, the stationary phase, flocculation, and finally dormancy (*Figure 1.1*).

The lag phase commences when the yeast is pitched into aerated wort. Yeast metabolism is slow at first, as the cells prepare enzyme systems necessary for growth. The yeast absorbs wort oxygen to "bank" as a raw material for cell membrane production during the growth phase. Brewer's yeast does not use this oxygen for respiration when growing in nutrient-rich wort. The lag phase can last 6–20 hours, depending on yeast health, pitch rate, and wort temperature.

During the growth phase, the yeast utilizes simple wort nutrients such as sugars, amino acids, vitamins, and minerals. Wort carbohydrate composition is particularly important because yeast cells prefer to maximize energy yields by using the simplest sugars first. Sucrose is broken down into glucose and fructose by an enzyme secreted by the yeast. These two sugars are utilized first, followed by maltose, the most abundant wort sugar, and finally by maltotriose. Glycogen is produced by the yeast during the active phase of fermentation. Brewer's yeast is unable to metabolize more complex sugars. Yeast growth results in a three- to more than sixfold increase in cell number during fermentation. It is during the

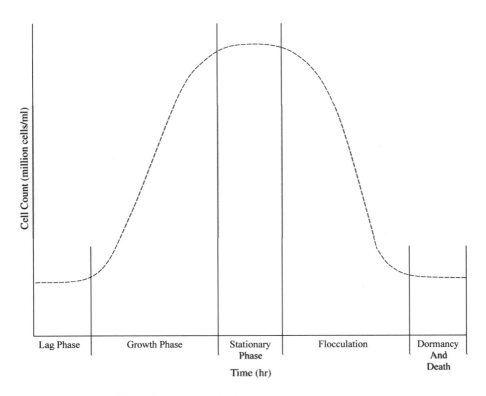

Figure 1.1. Yeast life cycle and growth phases.

growth phase that most of beer flavor, alcohol, carbon dioxide, and heat are produced. Nongrowing yeast metabolizes very slowly, resulting in a "stuck" or "hanging" fermentation. Healthy yeast growth, therefore, is the key to proper fermentation. *Figure 1.2* shows the essential mechanisms of yeast metabolism. As the wort nutrients are depleted, the growth phase comes to an end. The yeast has reached a peak cell count, and although fermentation continues at a slower rate, there is no more growth. The yeast enters the stationary phase.

As the usable nutrients are depleted and toxic compounds such as alcohol and carbon dioxide build up, the yeast prepares for life in a hostile environment. During the stationary phase, the yeast utilizes storage compounds such as glycogen, which will be consumed to sustain life during dormancy. The yeast flocculates and is harvested from the fermenter. If the yeast is not harvested promptly, it begins to die and releases enzymes and other cell materials to the detriment of beer quality. This last process is called autolysis.

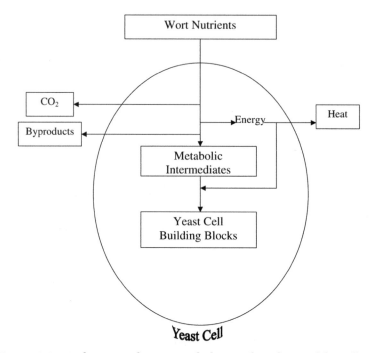

Figure 1.2. Mechanisms of yeast metabolism and workings of the cell.

14. How much alcohol, carbon dioxide, and heat are produced by fermentation?

One pound of fermented wort sugar can yield over 0.50 pounds of ethanol, about 0.46 pounds of carbon dioxide, and 252 BTU (586.6 kJ/kg) of heat. A barrel (1 bbl) of ale at high kräusen might drop 4°P (apparent extract) in 24 hours, producing about 1.4 standard cubic feet (scf)/bbl-hr (0.034 m³/hl-hr) of carbon dioxide and 92 BTU/bbl-hr (82 kJ/hl-hr) of heat.

Wort Composition and Fermentation Biochemistry

15. What is brewer's wort?

Wort is a liquid extract of malts, cereal grains, and hops produced in the brewhouse. It is the raw ingredient for beer fermentation. All-malt wort normally contains sufficient nutrients for proper fermentation, yeast health, and, hence, beer flavor. Three limiting factors to consider, however, are free amino nitrogen, zinc, and dissolved oxygen.

16. What are amino acids?

Amino acids are the basic building blocks of all proteins. Twenty different amino acids can be combined by yeast to create all the proteins necessary for enzymes and cell structures. Although all of these amino acids may be present in wort, yeast prefers to build new amino acids during fermentation.

17. Why does yeast manufacture amino acids when wort already contains these compounds?

It takes less energy for the yeast to manufacture new amino acids than to absorb and "inventory" all 20 types of wort amino acids at once. Should there be a shortage of any amino acid types in wort, the yeast must be able to create them in order to survive.

18. What is free amino nitrogen?

Free amino nitrogen (FAN) refers to the nitrogenous compounds, including amino acids (excluding proline) and ammonia, in wort or beer. Yeast needs this nitrogen for proper growth. Therefore, wort FAN levels have a great impact on beer flavor. FAN measurement is based on the ninhydrin reaction and is expressed as milligrams of FAN per liter or as milligrams per 100 ml.

FAN is a general measure of yeast-assimilable nitrogen, which is the aggregation of the individual wort amino acids and small peptides. Wort FAN content is an important factor in the formation of new amino acids by yeast, synthesis of new structural and enzymatic proteins, yeast viability and vitality, fermentation rate, ethanol tolerance, carbohydrate uptake, and attenuation. The practical brewer must be cautioned, however, that a given level of FAN does not define the actual amino acid makeup of the wort and that it is, in fact, the actual combinations and amounts of individual amino acids that affect fermentation and beer flavor. Nonetheless, healthy fermentation requires a proper level of FAN. Also, the difference between the FAN levels in wort and final beer can be used as a measure of fermentation vigor.

19. What is yeast-assimilable nitrogen?

Yeast-assimilable nitrogen is the nitrogen that the yeast cell can take up during fermentation, expressed in milligrams per liter. It includes amino acids, ammonia, and very small peptides.

20. What is total nitrogen?

Total nitrogen is all of the nitrogen in the wort, including amino acids, ammonia, peptides, proteins, nucleic acids, vitamins, etc.

21. What are typical FAN levels in wort and beer?

Wort FAN should be greater than 15 mg/100 ml but lower than 35 mg/100 ml. The optimum for lager beer worts is 21–23 mg/100 ml. Some brewers have observed that a ratio of FAN to total wort nitrogen of 21–22% is important for proper fermentation rather than the absolute level of FAN. FAN levels normally drop by 5–7 mg/100 ml during fermentation. In finished beer, FAN levels generally fall between 8 and 14 mg/100 ml in all-malt beer, as low as 5 mg/100 ml in adjunct beers, and as high as 29 mg/100 ml in doppelbocks. FAN levels in ales and weiss beer are usually on the low side of average, while lagers are on the high side.

22. How are total nitrogen and FAN measured?

To determine FAN levels, refer to ASBC *Methods of Analysis,* Wort-12, or MEBAK *Brautechnische Analysenmethoden,* Band II, 2.8.4.1. The total nitrogen content has historically been measured by the Kjeldahl method. Presently, the Dumas method is preferred. This number is not the same as yeast-assimilable nitrogen.

In a small brewery, it is not necessary for the brewer to test for total nitrogen or FAN unless deficiencies in fermentation are noted. However, it may be beneficial to measure total nitrogen and FAN in wort and beer, for example, once per year as a benchmark. Outside laboratories can be contracted to perform these tests, because they may be beyond the scope of most small brewery laboratories. In a large specialty brewery with a spectrophotometer, however, the measurement of FAN levels is simple to perform, and the brewer might find benefit in tracking FAN levels in wort, fermented beer, and final beer monthly or even weekly.

23. What effect does the FAN level have on fermentation?

High FAN levels promote excessive yeast growth, high diacetyl levels, high beer pH, and susceptibility to bacterial attack. Low FAN levels in wort can lead to sluggish fermentation and high levels of diacetyl in the final beer.

24. Can brewers control wort FAN levels?

When brewing all-malt beers, excessive wort FAN levels are best dealt with by choosing malt with less than 11.5% protein and moderate modification (e.g., S/T ratio [soluble protein as a percent of total protein] of 39–44%). In fact, FAN is a more important parameter to note than the S/T ratio when considering fermentation health. Starting the mash temperature at or above 144°F (62°C) also helps minimize wort FAN levels.

Low FAN levels are usually a problem only when adjuncts contribute more than 40% of wort extract. Most of wort FAN is produced during malting, but the brewer can slightly increase FAN levels by using a thick mash (e.g., 3:1), mashing at pH <5.5, and including a mash rest between 98 and 143°F (37–62°C). However, when modern malt varieties are used, protein rests longer than 15 minutes can have negative effects on beer quality. It should also be noted that the maltster can manipulate the malting process to control the level of wort FAN, but the barley variety and the geography of the growing area are responsible for the amino acid composition of the FAN (X. Yin, *unpublished data,* 2004).

25. Why is zinc important to fermentation?

Yeast needs zinc for proper growth and fermentation. It is especially important in the enzyme ethanol dehydrogenase (a zinc-metalloprotein), which is necessary for the production of ethanol. Most of the vitamins, minerals, and other micronutrients needed by yeast are supplied by an all-malt wort. However, wort can be deficient in zinc, and some of the zinc that is present can be bound up in natural chelating compounds, making the mineral unavailable to the yeast. In fact, most of the zinc dissolved into the wort from malt, hops, and water is lost in the spent grain and trub.

26. How are wort zinc levels measured?

Measurement of wort zinc requires an atomic absorption (AA) spectrophotometer or an inductively coupled plasma (ICP) spectrophotometer. This analysis can be contracted to an outside laboratory.

27. How much zinc should wort contain?

The zinc content of cold wort should be at least 0.08 ppm. Some beers may benefit from levels as high as 0.30 ppm.

28. How can brewers influence zinc levels in wort?

Zinc occurs naturally in malt, and its extraction can be enhanced by using highly modified malt, extending the protein rest, using a thin mash (e.g., 4:1), and mashing at pH 5.0–5.2 (e.g., using sour wort). Unfortunately, manipulation of the mash will have only a minimal effect on the final wort zinc level and may have other negative effects on beer quality. Overzealous use of kettle coagulants sometimes results in a very clear, cold wort lacking in zinc as well as other vital yeast nutrients.

Various yeast nutrient powders and yeast extracts that include zinc in their formulations are commercially available and can be added to the kettle or whirlpool. Brewers sometimes add food-grade zinc sulfate or zinc chloride to the brew kettle before knockout. About 50% of zinc is lost in the settled hot and cold trub, so a kettle addition of 0.055 g of food-grade zinc sulfate monohydrate (also called zinc sulfate powder) per hectoliter, 0.088 g of food-grade zinc sulfate heptahydrate per hectoliter, or 0.042 g of food-grade zinc chloride per hectoliter results in about 0.1 ppm zinc in the cold wort. Nonetheless, it is best for the brewer to avoid the use of yeast nutrients or added zinc unless a deficiency has been documented by laboratory analysis or the yeast shows poor fermentation quality.

In order to avoid the use of a non-*Rheinheitsgebot* additive, some brewers include zinc-containing metals in their process equipment. For example, galvanized water piping or even sacrificial bronze shoes that contain zinc on lauter tun rakes are used.

Overuse of zinc can result in a very rapid fermentation and off-flavors and may cause premature yeast flocculation. Gross overuse can be toxic to yeast. Added zinc absorbed by the yeast will influence subsequent fermentations. Therefore, some brewers add zinc during yeast propagation only. Yeast strains vary in their requirements for zinc and thus react differently to additions.

29. What are food-grade additives?

Any additive used in the brewing process not only must be listed as acceptable for use in the brewing process but must also meet standards for food-grade materials. Food-grade chemicals are labeled USP grade (United States Pharmacopoeia), NF grade (National Formulary), or FCC grade (Food Chemical Codex). International equivalents are the EP (European Pharmacopoeia), BP (British Pharmacopoeia), and JP (Japanese Pharmacopoeia).

30. Why is oxygen needed in cold wort?

Yeast uses oxygen dissolved in wort to synthesize unsaturated fatty acids and sterols, which are essential for cell membrane development. A sufficient amount of dissolved oxygen (DO) ensures high yeast growth, a quick start of fermentation, and good maturation. Brewer's yeast growing in wort does not use oxygen for respiration.

31. When should wort be aerated?

Most brewers add air or oxygen to wort as it exits the wort cooler. Once fermentation starts, ideally, no further air or oxygen should be added. Some brewers may need to rouse or add air to high-gravity beers and some ales during fermentation, but this practice may lead to problems with flavor stability.

Although it is not a common practice, some brewers inject air into hot wort as it enters the wort cooler instead of after it has cooled. The high wort temperature offers some protection from microbial growth from the air or aeration device and promotes superior mixing of the oxygen and wort as they travel through the wort cooler. However, this method is not recommended, because oxidation reactions occur at a rapid rate and lower gas solubility at such high temperatures.

32. What devices are used to dissolve air or oxygen in wort?

Equipment used for wort aeration includes stainless steel or ceramic sintered stones (*Figure 1.3*), static mixers, two-component jets (*Figures 1.4* and *1.5*), and venturi pipes (*Figure 1.6*).

33. How much oxygen is needed for proper fermentation?

The oxygen requirement for different yeast strains varies. However, typical yeast requires a DO content of at least 5 ppm, and the best results are achieved with 8–10 ppm in standard worts. A rule of thumb for high-gravity worts is 1 ppm per degree Plato. Therefore, high-gravity worts require a DO level well above the air saturation level, which is no more than 10 ppm in wort. In such cases, pure oxygen or a blend of air and oxygen is needed for proper aeration.

34. How is dissolved oxygen measured in wort?

It is difficult to properly measure wort DO, especially after yeast is pitched, so many brewers concentrate on ensuring a consistent air or oxygen flow rate into the wort. This flow rate is normally based on empiri-

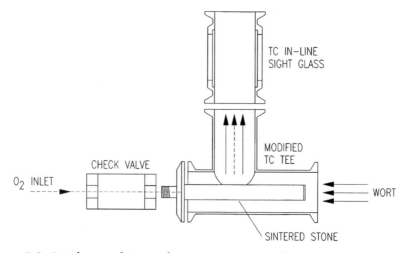

Figure 1.3. Stainless steel sintered stone arrangement for wort aeration. TC = triclamp. (Courtesy of JV NorthWest)

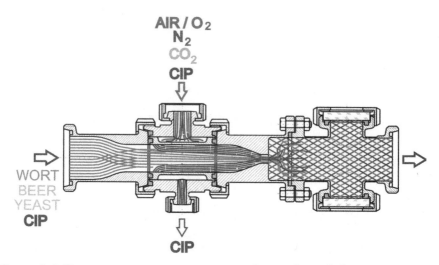

Figure 1.4. Two-component jet wort aerator with gas inlet and clean-in-place (CIP) capabilities. (Courtesy of Essau-Hueber, GmbH; all rights reserved)

cal results. In other words, as long as a consistent flow of gas is guaranteed and fermentation rates and beer flavor are satisfactory, it is assumed that the DO level is correct for a given yeast strain, beer type, and plant design.

However, wort DO can be measured at the inlet to the fermentation tank or at the fermenter zwickel with an inline or portable DO meter during the wort-cooling process (*Figure 1.7*).

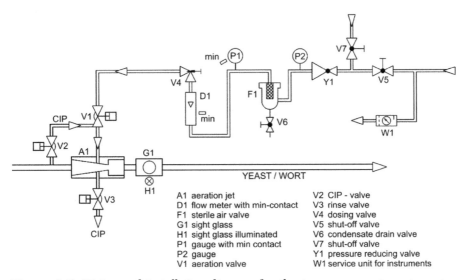

Figure 1.5. Piping and installation diagram for the two-component wort aerator. (Courtesy of Essau-Hueber, GmbH;)

It is best to take this measurement before yeast is pitched, but a reasonable estimation can still be made if the measurement is taken during, but not after, yeast is added to the wort. Active yeast, once in contact with the wort, quickly absorbs the DO. The brewer should also be aware that in-line measurements of wort DO may yield a false high reading if the pressure in the wort piping is substantially higher than that in the fermenter or if air gas bubbles are present. As the pressurized, oxygen-saturated wort enters the lower-pressure environment of the fermenter, it may degas slightly.

35. Do all brewers inject air or oxygen into wort?

Almost all brewers add compressed air or bottled pure oxygen to their cold wort. Brewers who do not inject air at all rely instead on atmospheric air pickup, e.g., from agitation by filling an open fermenter from the top through a fish-tail spray. In general, this method may achieve adequate aeration (which appears to work with some ale strains) if open fermenters and open yeast brinks are used, but it does not dissolve sufficient oxygen when most yeast strains or high-gravity worts are used. Before the advent of compressed air, this method was universally used with good results because it was combined with the use of a coolship, open wort cooler, open yeast screening, open yeast storage, and open fermentation and/or rousing, which all contributed to the total DO.

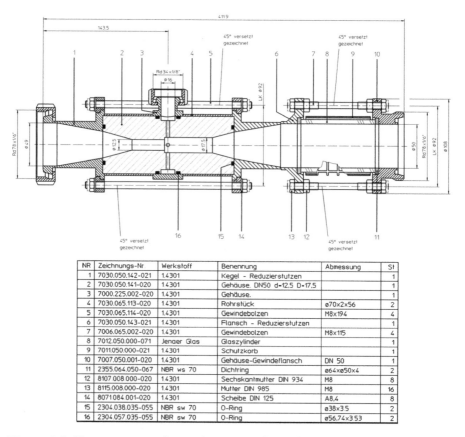

NR	Zeichnungs-Nr	Werkstoff	Benennung	Abmessung	St
1	7030.050.142-021	1.4301	Kegel - Reduzierstutzen		1
2	7030.050.141-020	1.4301	Gehäuse. DN50 d=12,5 D=17,5		1
3	7000.225.002-020	1.4301	Gehäuse.		1
4	7030.065.113-020	1.4301	Rohrstück	ø70x2x56	2
5	7030.065.114-020	1.4301	Gewindebolzen	M8x194	4
6	7030.050.143-021	1.4301	Flansch - Reduzierstutzen		1
7	7006.065.002-020	1.4301	Gewindebolzen	M8x115	4
8	7012.050.000-071	Jenaer Glas	Glaszylinder		1
9	7011.050.000-021	1.4301	Schutzkorb		1
10	7007.050.001-020	1.4301	Gehäuse-Gewindeflansch	DN 50	1
11	2355.064.050-067	NBR ws 70	Dichtring	ø64xø50x4	2
12	8107.008.000-020	1.4301	Sechskantmutter DIN 934	M8	8
13	8115.008.000-020	1.4301	Mutter DIN 985	M8	16
14	8071.084.001-020	1.4301	Scheibe DIN 125	A8,4	8
15	2304.038.035-055	NBR sw 70	O-Ring	ø38x3,5	2
16	2304.057.035-055	NBR sw 70	O-Ring	ø56,74x3,53	2

Figure 1.6. Venturi aeration device. (Courtesy of Kieselman GmbH)

A coolship is a large, shallow, open-topped vessel used before the advent of the whirlpool. The large surface area and shallow depth allowed for efficient separation of hot break and some natural wort aeration, as well as cooling.

36. What factors affect the actual gas flow rate needed to aerate wort?

a. Aeration equipment design. Depending on the method of injection, some of the gas will not bond and will rise out of the top of the fermenter.

b. Gas-delivery pressures. Gas-supply pressure must be greater than the pressure in the wort line and typically is 3–4 bar.

c. Piping and fermenter back-pressure. The length of the wort line, the amount of turbulence in the line, system pressure, and fermenter

Figure 1.7. Sampling for dissolved oxygen (DO). Note zwickel with DO meter on tee attached to tank inlet. (Photo by Cynthia Stalker)

fill height influence the amount of oxygen dissolved in the wort and the amount that remains in the wort once it reaches the fermenter. Wort can become "supersaturated" in a high-pressure wort line, but this excess oxygen can flash out of solution once the wort enters the lower-pressure environment of the fermenter.

d. Oxygen source. Pure oxygen is 4.75 times more soluble than air in wort, because air contains only 21% oxygen by volume.

e. Wort temperature. Air and oxygen are less soluble in warm wort.

f. Wort gravity. Air and oxygen are less soluble in high-gravity wort.

g. Brewery elevation. Air and oxygen are less soluble in wort at high elevations.

h. Amount of natural absorption. Wort may naturally pick up a small amount of oxygen during the wort cooling process. In a closed system from whirlpool to fermenter, there is very little air pickup; but if an open starter tank is used, the DO in wort may increase by more than 1 ppm from exposure to the atmosphere.

i. Target oxygen-saturation level. As the oxygen level in the wort approaches saturation, it becomes difficult to dissolve additional oxygen. When using compressed air, brewers aim for nearly 100% saturation (O_2 level of 8–10 ppm). Approximately 40% more gas flow may be required to achieve 100% saturation than to achieve 90%.

37. How can the saturation DO be estimated for a given wort?

The solubility of oxygen in wort is influenced by the gas type (air or pure oxygen), wort gravity, wort temperature, brewery elevation, and fermenter depth.

The following equations can be used to estimate wort saturation DO:

Saturation DO from compressed air = 9.4 ppm $\times g \times t \times e \times p$

Saturation DO from bottled oxygen = 44.2 ppm $\times g \times t \times e \times p$

where

g = correction factor for wort gravity (°P)

t = correction factor for wort temperature (°C)

e = correction factor for brewery elevation (feet)

p = correction factor for hydrostatic fermenter pressure

Use *Table 1.1* to find g, t, and e. Calculate p as follows:

$$p = \frac{[h\,(\text{ft}) \times 0.5 \times 0.433\ \text{psi/ft}] + P\,(\text{psi}) + 14.7\ \text{psi}}{14.7\ \text{psi}}$$

where

h = fermenter height, in feet

P = fermenter head pressure, in psi

Table 1.1. Factors for calculating dissolved oxygen levels in wort

g (Wort gravity)		t (Wort temperature)		e (Brewery elevation)	
°P	Factor	°C	Factor	Feet	Factor
10	1.04	5	1.10	Sea level	1.000
12	1.00	10	1.00	1,000	0.964
14	0.97	15	0.88	2,000	0.930
16	0.93	20	0.80	3,000	0.897
18	0.88	30	0.70	4,000	0.865
				5,000	0.834
				6,000	0.804

Example.   Calculate the saturation DO for a 12°P wort cooled to 68°F (20°C) and aerated with oxygen. The wort in the fermenter is 12 ft deep and at atmospheric pressure in a brewery at an elevation of 4,000 ft.

$$g = 1.00, \qquad t = 0.80, \qquad e = 0.865$$

$$p = \frac{(12 \text{ ft} \times 0.5 \times 0.433 \text{ psi/ft}) + 0 \text{ psi} + 14.7 \text{ psi}}{14.7 \text{ psi}} = 1.18$$

Saturation DO = 44.2 ppm × 1.00 × 0.80 × 0.865 × 1.18 = 36.1 ppm.

38. Is it possible to "overaerate" wort?

Many brewers use bottled oxygen instead of compressed air to aerate wort. Overaeration is a concern when pure oxygen is used, because DO levels greater than 40 ppm can be achieved. DO at these levels stimulates too much yeast growth and can have a negative effect on beer flavor (e.g., excessive higher alcohols and vicinal diketones) and yeast health. Excessive aeration with pure oxygen can even "poison" brewer's yeast, thus also affecting fermentation vigor. If bottled oxygen is used for aeration, the practical brewer might consider this gas the "fifth ingredient." By using oxygen instead of air, one can adjust the wort DO level from 0 to more than 40 ppm! The beer flavor can be significantly altered by control in such a large range.

When air is used as an oxygen source, it is generally not possible to overaerate wort. Air saturation of wort at 12°P and 68°F (20°C) is less than 8 ppm. However, when compressed air is coupled with an inefficient sintered stone, another problem exists. A high gas flow rate is required and overfoaming in the fermenter can occur. Not only does overfoaming cause a mess in the cellar and tank top fittings, but foam-positive agents are lost from the wort.

39. What is the theoretical amount of air or oxygen needed to aerate wort?

When using compressed air:

> 0.014 scf of compressed air/wort bbl (0.34 L/hl)
>   = 1 ppm DO
> 0.11 scf of compressed air/wort bbl (2.7 L/hl)
>   = 8 ppm DO

When using pure oxygen:

> 0.003 scf of pure oxygen/wort bbl (0.072 L/hl)
>   = 1 ppm DO
> 0.023 scf of pure oxygen/wort bbl (0.57 L/hl)
>   = 8 ppm DO

40. How is the actual aeration rate estimated?

The actual required gas flow rate to properly aerate wort is defined as the theoretical required flow rate multiplied by an equipment efficiency factor. The equipment efficiency factor is defined as

$$\frac{\text{Amount of gas delivered to the wort}}{\text{Amount of gas dissolved in the wort}}$$

The equipment efficiency factor is influenced by the type of aeration device used, type of gas used (air or oxygen), gas delivery pressure, and wort line hydraulics. *Table 1.2* estimates the equipment efficiency factors for both sintered stone-type and venturi-type aeration devices. Using this table and the theoretical gas flow rate as calculated above, the actual oxygen or airflow rate required can be estimated. Of course, the target DO cannot physically exceed the saturation DO as determined above.

> Wort flow rate (bbl/min)
>   × Theoretical gas flow rate to achieve 1 ppm DO (scf/bbl/ppm O_2)
>   × Required final DO (ppm O_2)
>   × Equipment efficiency factor (from *Table 1.2*)
>   = Actual required gas flow rate (scf/min)

Example.   Estimate the necessary gas flow rate to dissolve 8 ppm oxygen into wort using pure oxygen and a sintered stone, cooling 15 bbl in 45 minutes. The equipment efficiency factor is estimated to be between 1.5 and 4. Hence, the low-end estimated flow rate is as follows:

Table 1.2. Estimated gas flow rate efficiency factors

Equipment design	Compressed air	Pure oxygen
Sintered stone or gas sparger (0.5 micron)	2–12	1.5–4
Venturi system	1.25–4	1–2.25

15 bbl/45 min × 0.003 scf/bbl/ppm × 8 ppm

    × 1.5 (Efficiency factor for sintered stone using pure oxygen)

    = 0.012 scf/min

The high-end estimated flow rate is as follows:

15 bbl/45 min × 0.003 scf/bbl/ppm × 8 ppm × 4 = 0.032 scf/min

To convert to liters per minute, multiply by 28.32 normal liters (NL)/ft^3.

Therefore, in the above example an estimated oxygen flow rate of 0.012–0.032 scf/min (0.34–0.91 NL/min) will be required to achieve a DO level of 8 ppm. This is only an estimated oxygen flow rate. Because there are so many variables, a DO meter should be used to determine the actual final DO.

41. What are the standard cubic foot, normal cubic meter, and normal liter?

Gases expand or contract depending on their absolute temperature and pressure. If the pressure increases or the temperature decreases, the mass of gas in 1 ft^3 will increase. To avoid confusion when specifying an amount of gas, the working pressure and temperature must be defined, in addition to the volume. A standard cubic foot (scf) of gas is measured at a standard temperature and atmospheric pressure (STP), defined as 32°C and 14.7 psi absolute (0°C and 1.0 bar absolute). The metric counterparts of the standard cubic foot are the normal cubic meter (Nm^3) and the normal liter (NL):

$$1 \text{ scf} = 28.32 \text{ NL} = 0.02832 \text{ Nm}^3$$

Many brewers use a combination pressure regulator and flowmeter to inject gas into wort. Most flowmeters are calibrated to deliver air or oxygen in standard cubic feet per minute or normal liters per minute at standard temperature and pressure. Gas supply pressure is normally 40–50 psi in order to ensure proper flow of gas into the wort stream.

42. How much does a standard cubic foot of oxygen or air weigh?

a. Air at STP weighs 0.080 pound (lb)/ft³ (1.29 g/L). At 68°F and atmospheric pressure, air weighs 0.075 lb/ft³ (1.20 g/L).

b. Oxygen at STP weighs 0.089 lb/ft³ (1.43 g/L). At 68°F and atmospheric pressure, oxygen weighs 0.082 lb/ft³ (1.31 g/L).

43. How can the mass of delivered gas be useful to brewers?

This information is important if a brewer uses a mass flowmeter, or scale, instead of a volume flowmeter. In the example above, in a 15-bbl brew aerated to a DO level of 8 ppm, the mass flow rate and total mass delivered for the high-end estimate is calculated as follows.

What is the mass flow rate for the above example?

$$0.032 \text{ scf of oxygen/min} \times 0.089 \text{ lb of oxygen/scf}$$
$$= 0.0028 \text{ lb of oxygen/min}$$

or, in the metric system,

$$0.91 \text{ NL of oxygen/min} \times 1.43 \text{ g of oxygen/NL}$$
$$= 1.3 \text{ g of oxygen/min}$$

What is the total mass of oxygen delivered?

$$0.0028 \text{ lb/min} \times 45 \text{ min} = 0.126 \text{ lb of oxygen}$$

or, in the metric system,

$$1.3 \text{ g of oxygen/min} \times 45 \text{ min} = 59 \text{ g of oxygen}$$

44. What are the by-products of yeast growth?

The flavor of beer is impacted not only by the type of raw materials used but also by the numerous products of yeast metabolism. One has only to taste wort side by side with finished beer to appreciate this fact. Important products of yeast activity include organic acids, fatty acids, ketones, higher alcohols, esters, sulfur compounds, and acetaldehyde. Acetaldehydes, certain ketones, and some sulfur compounds are considered off-flavors in finished beer. Fortunately, unlike the other classes of compounds, they are reduced by yeast in late fermentation and aging. See *Figure 1.8.*

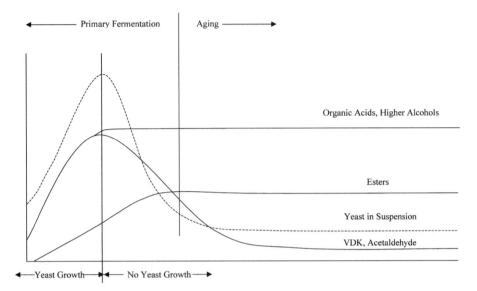

Figure 1.8. By-product formation over the fermentation and aging cycles, showing the reduction of certain products post-fermentation. VDK = vicinal diketones.

45. What are organic acids in beer?

More than 100 organic acids have been identified in beer. Some acids come directly from the raw materials or brewhouse operations, but most are formed by the yeast during the growth phase. These acids are intermediary compounds produced by yeast during amino acid synthesis. Excess acids are excreted by the yeast in order to regulate the cell's internal metabolic balance. Some of these excreted acids, such as acetic acid, are reabsorbed by the yeast for continued metabolism later in fermentation. See *Figures 1.2* and *1.9.*

46. Do organic acids affect beer pH?

During fermentation, the pH drops from about 5.25 ± 0.25 to 4.25 ± 0.35. Organic acids (such as succinate, lactate, pyruvate, and malic) are only partly responsible for this effect.

47. What other factors affect beer pH?

This pH drop during fermentation is not completely understood, but some of the other factors include absorption of basic amino acids and phosphates coupled with production of carbon dioxide and hydrogen ions by the yeast during metabolism.

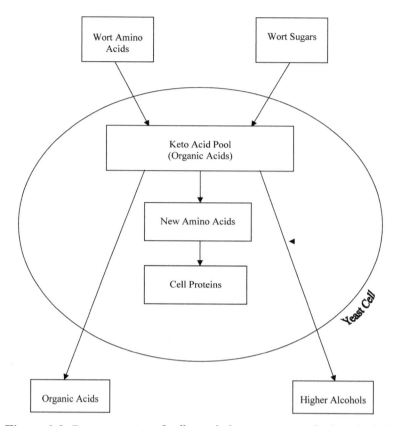

Figure 1.9. Representation of cell metabolism concerning higher alcohols and organic acids.

48. Why is the drop in pH critical to beer quality?

The acids produced by fermentation have an effect on the liveliness and drinkability of beer. As pH drops during fermentation, the beer color (SRM value) also decreases. (SRM is a color intensity scale based on absorbance.) A pH between 4.0 and 4.5 also helps the beer resist bacterial contamination, increases beer physical haze stability, and hastens the rate of maturation during aging.

The final beer pH has an effect on the flavor stability of the beer and on the type and formation of aged flavor compounds. Low beer pH levels are thought to favor the formation of compounds such as *trans*-2 nonenal or "cardboard" aromas; higher pH levels are more prone to favor caramel or sherry-type aromas.

Low final beer pH levels are generally associated with overall lower flavor stability. The most likely explanation is that, at beer pH less than

4.0, the superoxide radical is protonated, which renders it into its most damaging form (Dr. Charles Bamforth, *personal communication,* October 2004).

Brewers monitor pH during the course of fermentation. A rapid start to fermentation and quick drop in pH indicate a sound fermentation. During aging, the pH may increase slightly—especially at the bottom of the tank near the yeast layer—as cell autolysis begins. This negatively impacts beer flavor and foam quality.

49. What factors promote a favorable beer pH?

a. Lowering brewing water alkalinity

b. Lowering wort pH with the addition of sour wort or food-grade acid. (Lowering the mash pH without additional treatment of the kettle wort can actually increase beer pH, as a result of improved mash enzyme action and release of buffering compounds.)

c. Using low protein malts or diluting malt with adjuncts

d. Using highly colored malts

e. Any factor that promotes yeast health and vigorous fermentation

f. Avoiding yeast autolysis by removing yeast immediately after flocculation and repitching as soon as possible

50. What are fatty acids?

Fatty acids are a class of organic acids associated with lipids and are an essential component of yeast cell membranes. They have medium or long nonpolar hydrocarbon tails and a polar carboxylic acid group.

Fatty acids in beer come from four sources:

a. Hops, especially stale hops

b. Trub carryover (a high quantity of hot break and less favorable whirlpool design foster carryover of trub into the fermenter)

c. Malt lipids, depending mainly on the lautering process

d. Yeast metabolism

High molecular weight fatty acids, such as linoleic acid, generally originate from the mash and are important nutrients for yeast growth and cell membrane development during fermentation. During fermentation, yeasts produce lower molecular weight fatty acids, such as caproic (hexanoic), caprylic (octanoic), and capric (decanoic) acids, which are building blocks of cell membrane materials. These compounds are not a problem during fermentation but can contribute a strong "soapy," "goaty" character and cause a decrease in beer foam stability in aged beers when released through yeast autolysis. Warm or excessive aging of beer results in elevated levels of these fatty acids and their esters. It is important to get the yeast off the beer once primary fermentation is complete. When using the unitank system, the yeast should be gently bled from the tank cone every two to three days early in maturation and once per week later in aging.

51. Is there a simple method to test for the extent of yeast autolysis?

After the yeast slurry is harvested from the fermenter, centrifuge a 50-ml sample at 1,000 rpm for 5 minutes. Measure the pH of the supernatant. If it is 0.5 pH units or more above the beer pH, significant autolysis may have occurred.

An increase in beer FAN or fatty acids during beer aging also indicates excessive yeast degradation. A fatty acid, capric acid (decanoic), in beer can be used as an indicator compound for yeast autolysis and should be 0.5 ppm or less. The combined total of the main fatty acid esters (ethyl hexanoate, ethyl octanoate, and ethyl decanoate) should be less than 0.35 ppm.

52. Which ketones are most important in beer?

The main ketones of interest to brewers are 2,3-butanedione (diacetyl) and 2,3-pentanedione. These ketones contain two adjacent carbonyl groups and therefore are referred to as vicinal diketones (VDKs). See *Figure 1.10.*

VDK is a by-product of protein anabolism. In order to grow, yeast cells manufacture amino acids from wort sugars and FAN. Diacetyl and pentanedione are by-products of this metabolism. During vigorous fermentation, yeast cells produce an "inventory" of these building blocks and release them into the beer. As available nutrients become scarce later in fermentation, the yeast cells reabsorb this "inventory" to use for continued metabolism. Late in a healthy fermentation, the yeast will use most of the excess VDK. If the yeast has produced too much "inventory," or is not healthy enough to continue strong metabolism, the final beer may contain excessive VDK. Therefore, VDK is an indicator compound for faulty or incomplete fermentation.

Recent research has shown that a small amount of VDK can be produced during Maillard reactions, which are significant during malt kilning, brewhouse operations, and beer staling (Coghe et al., 2004).

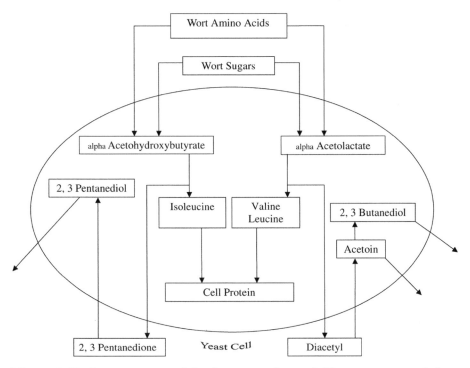

Figure 1.10. Representation of the formation of vicinal diketone compounds by a yeast cell.

Some bacteria, especially *Pediococcus* strains, can also produce large amounts of diacetyl.

53. What is the flavor of VDK?

Diacetyl and pentanedione have similar tastes, described as "buttery," "butterscotch," "honey," or "milky." High VDK levels interfere with beer drinkability and are considered to be a significant off-flavor.

54. Why should brewers monitor VDK levels?

During fermentation, VDK levels peak at the same time that yeast growth finishes. Therefore, brewers like to see peak yeast cell count and VDK as early as possible in the fermentation so that there is sufficient time for reduction. This is accomplished by pitching vigorous yeast at the correct rate into well-aerated, properly made wort at a moderate temperature.

Measuring VDK late in fermentation and during aging helps the brewer decide when to cool the fermenter and when to filter.

55. At what point should brewers measure VDK?

Every fermenter VDK level should be checked at the end of fermentation before the beer is cooled to aging temperature. Beer ready for filtration and packaging can also be checked. Peak VDK levels during main fermentation need to be checked only if problems arise.

56. How can brewers measure VDK?

Three methods of measuring VDK are available:

a. The simplest method is a go/no-go test suitable for small producers. Collect two 200-ml samples of beer in two 500-ml flasks. Cover both loosely with foil. Store one flask at room temperature. Gently warm the second flask in a bath of hot water at 140–160°F for 10 minutes. Heating the beer promotes the rapid oxidation of the relatively flavorless VDK precursors to VDK. A heating step is vital for the proper measurement of total VDK levels, which includes these precursors. Cool the heated sample back to room temperature using, for example, ice water or a cold water bath. Remove the foil covers, swirl the samples, and smell each. If no diacetyl is detected in either sample, it can be assumed that the level is below your taste threshold and the beer is ready to package. If the heated sample smells buttery but the cold sample does not, then the beer is progressing satisfactorily but needs further aging to reduce VDK and its precursors. If both the samples smell buttery, it may mean that the beer is still far too young to package and in need of further aging, that your yeast strain lacks the ability to properly reduce the VDK, or, in severe cases, that a bacteria infection is present (de Piro, 2004). One or more staff members must be familiar with and sensitive to diacetyl flavor. Periodic training by using a commercially available beer flavor standards kit or by attending off-flavor tasting seminars is helpful.

b. The second method requires a simple glass distillation apparatus and spectrophotometer. There are numerous distillation methods. A sound example can be found in MEBAK *Brautechnische Analysenmethoden*, Band II, Method 2.23 Vicinale Diketone. A distillation apparatus and this method written in English are available from Lg–automatic aps (Denmark) through their U.S. representative Profamo Inc. (Sarasota, Florida). Five samples can be analyzed in about 90 minutes.

c. The third method requires a gas chromatograph. While the second method measures total VDK, this method allows the brewer to differentiate between diacetyl and 2,3-pentanedione.

57. Can the brewer estimate the time and temperature needed for diacetyl reduction?

It is best to track VDK reduction by actual measurement. However, if such testing is beyond the scope of a small brewery, the following equation can be used to estimate when VDK reduction is sufficient and the beer can be cooled to aging temperature. Of course, this exercise will yield only a rough estimate, as the actual reduction of VDK is influenced by many factors. Nonetheless, this equation does illustrate the importance of time and temperature on VDK reduction (Heyse, 1994, pp. 184–186).

Days of diacetyl rest required

$$= \frac{100 - \sum(\text{Average daily ferment temperatures, in °C})}{\text{Diacetyl rest temperature, in °C}}$$

Example 1.   How long should a diacetyl rest at 68°F (20°C) last if primary fermentation takes three days at 68°F (20°C)?

$$\frac{100 - (20°C + 20°C + 20°C)}{20°C} = 2 \text{ days}$$

Example 2.   How long should a diacetyl rest at 52°F (11°C) last if primary fermentation starts at 48°F (9°C), rises in two days to 52°F (11°C), and is complete after six days?

$$\frac{100 - (9°C + 10°C + 11°C + 11°C + 11°C + 11°C)}{11°C} = 3.4 \text{ days}$$

Example 3.   How long should a diacetyl rest at 43°F (6°C) last if primary fermentation starts at 43°F (6°C), rises to 48°F (9°C) over three days, is held one more day, and is cooled over three days to 43°F (6°C)?

$$\frac{100 - (6°C + 7°C + 8°C + 9°C + 9°C + 8°C + 7°C + 6'C)}{6°C} = 6.6 \text{ days}$$

58. Can a brewer tell if VDK is from yeast or bacterial contamination?

In a well-designed modern brewery, the incidence of bacteria contamination should be low. Because many specialty brewers ship unpasteurized beer, contamination levels should normally be below detection threshold. However, eventually every brewer will face some sort of contamination issue. Lactic acid bacteria—*Pediococcus*, in particular—

can produce large amounts of diacetyl. Unlike yeast, however, they do not produce pentanedione. The use of a gas chromatograph allows the brewer to monitor the ratio of diacetyl to pentanedione. An elevation in this ratio can indicate that the metabolic by-products of lactic acid bacteria are present.

59. What factors influence VDK levels during fermentation and lagering?

a. Wort FAN.   FAN levels in wort should be kept between 15 and 35 g per 100 ml. Use adjuncts at 40% or less of total extract. For all-malt beers, choose malt with 12.5% protein or less. Avoid severely undermodified malts (e.g., less than 36% soluble nitrogen). Darker beers often contain higher VDK levels than pale beers, partly because dark malts contain less "starter sugars" and a different amino acid composition than pale malt.

b. Yeast strain.   Most yeast cells produce diacetyl at a similar rate, but they vary in their ability to reduce the same in late fermentation. Flocculent strains and respiratory-deficient cultures generally leave higher VDK levels.

c. Yeast health.   In order to ensure proper reduction of VDK, the yeast should be maintained in a vital state. Yeast should be harvested as soon as it flocculates and be stored at 32–39°F (0–4°C) for three days or less before repitching. Wort should be properly aerated at pitching.

d. Wort aeration.   Underaeration causes a sluggish fermentation, which can increase final VDK levels. Overaeration might encourage too much yeast growth, which promotes the production of VDK.

e. Pitch rate.   Underpitching can lead to higher final VDK levels.

f. Fermentation temperature.   Starting fermentation at a temperature slightly lower (such as 4°F lower) than that of peak fermentation can minimize diacetyl production. Letting fermentation temperature rise an additional 3–5 degrees F after the peak cell count has been achieved and maintaining this temperature for a short period at the end of primary fermentation help ensure a low final concentration of diacetyl.

g. Cell count in late fermentation.   The higher the amount of vigorous yeast cells in suspension late in fermentation, the quicker the reduction of VDK. At least 5 million cells per milliliter at the end of primary fermentation are required for proper secondary fermentation, although some brewers prefer 10 times that amount. The health of this yeast will ultimately determine the amount of yeast required.

h. Postfermentation air pickup.   If beer is pumped from a primary tank to an aging tank, air pickup should be minimized, as air will cause the yeast to produce more diacetyl.

i. Beer pH.   As pH decreases, the rate of VDK removal increases.

j. Kräusening.   Freshly fermenting wort added to the lager tank, particularly when flocculent yeast is used, can lower final VDK.

k. Lagering.   Diacetyl was not considered a significant problem in brewing until brewers started to shorten fermentation and aging times. A secondary fermentation with at least 5×10^6 cells per milliliter will lower VDK levels. The necessary aging time and temperature are dictated by the yeast strain.

60. Can beer with high diacetyl be "fixed"?

The brewer must first determine the reason for the high diacetyl level. If it is due to bacteria infection, the beer can be flash-pasteurized or sterile-filtered before reprocessing. However, in the case of bacterial infection, it is always best to avoid compounding the error; this is achieved by dumping the infected beer, stripping clean all exposed brewing equipment, and restarting production with freshly propagated yeast.

Assuming there is still sufficient viable yeast in the high-diacetyl beer (at least 5×10^6 cells per milliliter) and the high diacetyl is not due to bacterial contamination, the diacetyl level may come down with time in the cellar, but this might take weeks depending on the cellar temperature.

In order to reprocess high-diacetyl beer more quickly, the brewer can add kräusen beer to the problem batch. Kräusen beer is simply 18- to 24-hour-old, actively fermenting wort that has reached about 10–30% attenuation. The process for "repairing" high-diacetyl beer is as follows:

   a. If possible, allow the beer to warm slightly in the aging tank to a moderate temperature suitable for the particular yeast strain used.
   b. Add 10–15% (by volume) "kräusen" beer to the high-diacetyl beer.
   c. Continue maturation in the normal fashion. The active yeast from the kräusen beer can reduce the diacetyl level in the original beer in as little as one week.

Kräusening is discussed in more detail later in this chapter.

Of course, this process alters the flavor profile of the beer and is not guaranteed to work in all cases, especially if the kräusen beer is also defective. The brewery must decide whether it is reasonable to attempt to "save" the problem batch or dump it. In a specialty brewery, the decision is hopefully based on quality, not economic considerations.

61. What are higher alcohols in beer?

In addition to ethanol, at least 45 different alcohols have been identified in beer. These alcohols are referred to as higher alcohols, because their molecular weight is greater than that of ethanol. They make up one of the largest classes of flavor-active compounds in beer.

Beer higher alcohols are divided into two classes: aliphatic alcohols and aromatic alcohols.

62. What are aliphatic alcohols in beer?

The term *aliphatic* refers to organic compounds with open chains of carbon atoms. Aliphatic alcohols are volatile, straight-chained higher alcohols. They contribute a "hot," "solventy," or "alcoholic" taste and smell to beer. The most important aliphatic alcohols are *n*-propanol, *iso*-amyl alcohol (3-methyl butanol), active amyl alcohol (2-methyl butanol), and *iso*-butyl alcohol (2-methyl propanol).

63. What are aromatic alcohols in beer?

Aromatic alcohols are the second class of higher alcohols in beer. In spite of their name, aromatic alcohols are not strongly volatile. The term *aromatic* refers to organic compounds containing one or more six-carbon rings, a characteristic of benzene-based compounds. Aromatic alcohols contain phenols, which are benzene-based molecules. Although aromatic alcohols can be important in ale and weiss beer flavor, they are considered undesirable at high levels in lager beers.

The most important aromatic alcohol in beer is 2-phenylethanol. Moderate levels of 2-phenylethanol promote beer drinkability. 2-Phenylethanol is also an indicator compound for fermentation temperature. For example, the 2-phenylethanol content of a Pilsner fermented at 48°F (9°C) may be 10 ppm, while that of a Pilsner fermented at 54°F (12°C) may be more than 25 ppm. Generally, the 2-phenylethanol content should not exceed 30 ppm in lager and 45 ppm in ale.

64. Why are higher alcohols sometimes called fusel oils?

Fusel oils are the oily, low-volatile compounds left in a pot still after distillation. *Fusel* is a German word referring to a poor-quality distilled

Table 1.3. Higher alcohols commonly present in lager and ales

Compound	Aroma	Taste threshold (ppm)	Typical concentration (ppm)	
			Lager	Ale
Aliphatic alcohols				
n-Propanol	Alcoholic	600–800	6–17	20–45
iso-Amyl alcohol (3-methylbutanol)	Alcoholic	30–65	25–75	80–140
Active amyl alcohol (2-methylbutanol)	Alcoholic	50–70	10–20	9–41
iso-Butyl alcohol (2-methylpropanol)	Alcoholic	80–200	6–11	11–33
Aromatic alcohol				
2-Phenylethanol	Roses	40–125	4–32	8–50

spirit containing excessive levels of higher alcohols. It comes from an Old German word meaning "working poorly or in a sloppy manner"—in this case, referring to the production of low-quality spirits.·

65. How are higher alcohols measured in beer?

Higher alcohols are best measured by a gas chromatograph.

66. What higher alcohols are important in beer?

Table 1.3 shows some common higher alcohols and their concentrations as found in lagers and ales.

Although some of these compounds may be below threshold, they still can have an impact on beer flavor due to synergies with other compounds. Excessive levels of these alcohols can make beer taste empty and harsh. Total concentrations in lager beer are best kept below 100 ppm.

67. Why is there more than one type of amyl alcohol in beer?

There are actually eight different isomers of amyl alcohol. Only two, isoamyl and active amyl, are important in fermented products. The "active" form of amyl alcohol rotates polarized light. Active and isoamyl alcohols are together referred to as "combined amyl alcohols."

68. How are higher alcohols produced during fermentation?

Almost all higher alcohol production occurs in primary fermentation. Higher alcohols are by-products of amino acid synthesis. Yeast cells use existing wort amino acids and sugars to construct new amino acids. Organic acids are intermediary compounds in this process. Yeast cells are very efficient at dismantling and making amino acids and often produce an excess of organic acids. Yeast can convert the acids into alcohols and excrete them into the beer. Higher alcohols are therefore "overflow" compounds—they are not an end product of fermentation and, unlike diacetyl, are not reabsorbed and reused. See *Figure 1.9.*

69. What factors influence higher-alcohol production during fermentation?

a. Yeast strain.   The amounts of higher alcohols produced by different strains can vary by as much as fivefold.

b. Fermentation temperature.   Generally, higher fermentation temperatures produce greater levels of higher alcohols (particularly phenyl ethanol). Some yeast strains are more sensitive to this effect than others. Some brewers utilize top pressure during primary fermentation if conducting a rapid, warm fermentation to minimize this effect.

c. Pitch rate.   With high pitch rates, there is less yeast growth and, therefore, less higher-alcohol production. Underpitching can lead to excess levels.

d. Wort gravity.   Higher-gravity worts yield increased levels of higher alcohols.

e. Wort aeration.   Increased aeration rates or aeration after the start of fermentation leads to increased production of higher alcohols.

f. Wort composition.   Low FAN levels, particularly in high-adjunct beers, can lead to increased higher-alcohol production. Excessive levels of wort FAN or zinc can also increase higher-alcohol production.

g. Fermenter type.   Tall, slender vessels promote production of higher alcohols.

70. What are esters?

Esters are compounds made by combining an alcohol with an organic acid. More than 90 esters have been identified in beer. Because of their often strong, fruity character, they have a large impact on beer flavor.

71. How are ester levels measured?

Esters are measured using a gas chromatograph.

Table 1.4. Esters in beer

Compound	Aroma	Taste threshold (ppm)	Typical concentration (ppm)
Ethyl acetate	Solvent, nail polish	30	7–48
Isoamyl acetate	Banana	0.6–2	0.8–5.2
Ethyl caproate (ethyl hexanoate)	Apple, aniseed	0.2	0.05–0.3
Ethyl caprylate (ethyl octanoate)	Apple	0.3–1	0.1–0.5
Phenylethylacetate	Roses	3	0.1–1.5

72. What esters are most important in beer?

The most important esters in beer are ethyl acetate (solvent flavor), isoamyl acetate (banana flavor), phenylethyl acetate (honey flavor), ethyl caproate (ethyl hexanoate; apple flavor), and ethyl caprylate (ethyl octanoate; apple flavor).

73. What are ester levels in beer?

Although some of these compounds may be below threshold levels, they still can have an impact on beer flavor due to synergies with other compounds. *Table 1.4* gives some flavor characteristics and threshold values for esters commonly found in beer.

74. How are esters produced during fermentation?

Yeast ester biochemistry is poorly understood. Unlike most other yeast metabolic by-products, total ester levels peak after the growth phase ends. Late in fermentation, there is excess acetyl CoA, an important but acidic metabolic compound. It is believed that the yeast reduces acetyl CoA, as well as other alcohols and acids, by transforming them into esters, thus lessening their toxic effect. See *Figure 1.11*.

75. What factors influence ester production during fermentation?

a. Yeast strain.   Different strains produce different types and total amounts of esters.

b. Barley and malt.   Empirical evidence indicates that barley crop year, variety, and cultivation and harvest practices influence beer ester character. Malt overmodification can result in beer with decreased ethyl acetate concentrations.

c. Wort aeration.   Insufficient wort aeration increases final ester levels. Oxygen is needed by the yeast to make unsaturated lipids for cell

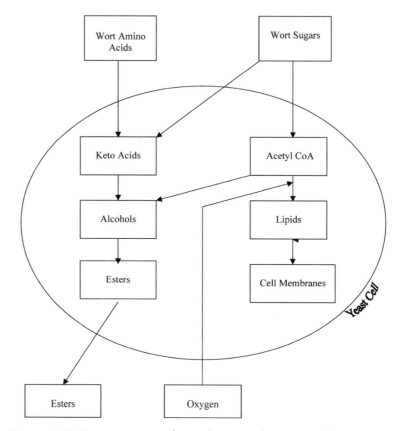

Figure 1.11. Representation of ester formation by a yeast cell.

membranes. If sufficient wort oxygen is available, the yeast can use acetyl CoA for lipid synthesis. Thus, late in fermentation, there will be less acetyl CoA that must be excreted as esters.

d. Wort sugar spectrum.   High levels of maltose in wort inhibit the transportation of esters out of the yeast cell. Worts produced with elevated glucose and lowered maltose concentrations result in beers with elevated levels of ester, especially isoamyl acetate (Herrmann et al., 2003).

e. Wort FAN levels.   Excess FAN levels (>22 mg/100 ml) or deficient FAN levels yield higher beer ester concentrations.

f. Cold wort lipid levels.   Hazy wort produces beer with lower ester levels.

g. Fermentation temperature.   Warm fermentation increases ester levels, although some yeast strains are less influenced by temperature than others. It is thought that yeast fermenting at higher temperatures needs less unsaturated lipid material, because its cell membranes are more "fluid" at the warmer temperature. Therefore, acetyl CoA would be in excess, leading to increased ester formation. Extremely cold primary fermentations can also lead to excessive ester production.

h. Pitch rate.   Lower pitch rates increase the production of some esters.

i. Original gravity.   Higher starting gravities can lead to more ester production. Higher wort aeration levels can help minimize this effect when brewing strong beers.

j. Fermentation vessel design.   Deep, closed fermenters lower ester levels, due primarily to higher hydrostatic pressure on the yeast, which represses alcohol and acetyl CoA production and subsequent ester formation. Shallow, open fermenters are preferred for the production of estery beers such as Bavarian-style weiss beers.

76. What is the significance of the ratio of higher alcohols to esters?

Although elevated ester levels are desirable in most ales and in some dark or amber lagers, lower levels are preferred in pale lagers. It has been observed by some brewers that the ratio of higher aliphatic alcohols to esters is important to beer flavor. These brewers believe that this ratio should be near 3:1 for maximum beer drinkability. Tall, narrow fermenters or excessive/repeated aeration can lead to a higher ratio. A beer with a ratio of 5:1 or more might taste thin and harsh, although this is by no means always the case.

77. What volatile sulfur compounds are important in fermentation?

Yeast metabolism produces volatile sulfur compounds like sulfur dioxide (SO_2), hydrogen sulfide (H_2S), and many organic compounds such as mercaptans and organic sulfides. These compounds are strongly aromatic and can affect the flavor of beer, even when present in parts per billion. The flavor threshold of SO_2 is 20 ppm, while the threshold for H_2S is only 5 ppb, and for some mercaptans as low as 0.5 ppb. Other sources of sulfur compounds in beer are malt, hops, processing aids, and microbiological contamination.

According to Bamforth, "Putrefaction; drains; garlic; burnt rubber; cat urine: are not exactly pleasant terms, yet they are the aroma descriptors typically associated with sulphur-containing compounds found in beer. It's a question of balance. When these materials are present in moderation, perfectly balanced against the myriad of other flavor compounds in beer, they can make a critical contribution to the quality of a product" (Bamforth, 2001).

Examples of two popular lagers, one brewed in Pennsylvania and the other in Munich, and an ale from Burton, England, all known for some sulfur characters, offer proof of this claim.

78. What is the significance of DMS in fermentation?

Dimethyl sulfide (DMS) is another important volatile sulfur compound. It has the aroma of corn above a threshold of 30–60 ppb and is associated with defects in malt quality, kettle boil, or extended whirlpool rests. DMS is formed from a malt-derived precursor, S-methyl methionine, when malt is kilned or wort is heated. Also, yeast cells are able to transform a second DMS precursor, dimethyl sulfoxide (DMSO), into active DMS. This process is strain dependent. Thirdly, DMS can be produced by microbiological contamination.

Like other volatile sulfur compounds, DMS will be scrubbed from solution during the kettle boil and in a vigorous fermentation. Some breweries go to great lengths to strip out DMS, utilizing air-stripping towers during wort cooling and during transfer from fermentation to lagering.

79. What are the other flavors of volatile sulfur compounds formed in fermentation?

Hydrogen sulfide has the aroma of rotten eggs. Sulfur dioxide has the aroma of burnt matches. Mercaptans have the aroma of rotten vegetables. Other organic sulfur compounds exhibit similar flavor characteristics. Over time, some of these flavors in finished beer will subside due mainly to oxidation.

80. Can brewers measure volatile sulfur compounds?

The measurement of volatile sulfur compounds is complicated. These flavors are often present at low concentrations and are prone to oxidation. Modern gas chromatography utilizing either flame photometric detectors or chemiluminescence detectors can be useful for detecting many sulfur compounds. Because most of the important compounds are strongly volatile, most specialty brewers simply rely on the nose of a qualified taster.

81. How are volatile sulfur compounds formed during fermentation?

Sulfur compounds are produced by yeast during its growth phase. Yeast utilizes sulfur-containing malt amino acids and sulfate ions from brewing water to create other sulfur-containing amino acids needed for yeast growth. Hydrogen sulfide and sulfur dioxide are by-products of this metabolism. Organic sulfur compounds are produced by similar pathways. Most yeasts (lager and weiss strains, in particular) can produce noticeable levels of volatile sulfur compounds early in fermentation, but healthy yeast will "mop up" excess levels late in fermentation and aging. High levels in finished beer can be due to a sluggish fermentation or to microbiological contamination.

Highly vital yeast utilizes more sulfate to form methionine and other sulfur-containing amino acids to build cell materials. Less-active yeast does not use the sulfate fully but reduces it to SO_2 and releases it into the fermenting wort.

82. What sulfur compounds are beneficial?

Some brewers use the significant antioxidant properties of naturally produced SO_2 to inhibit oxidation and prevent staling. By controlling pitch rate and limiting wort aeration to slightly stress the yeast, some strains can be manipulated to produce elevated SO_2 levels. In the United States, levels over 10 ppm, even from natural sources, require disclosure on the label, since some people can have severe allergic reactions to this compound. Some beers have, as part of their flavor profile, a noticeable sulfur character that in other brands may be perceived as a defect.

83. What factors influence the level of volatile sulfur compounds in fermentation?

a. Yeast strain. The amount of sulfur formed varies greatly among strains, and strain selection is one of the most important factors in control of many sulfur flavors. Ale yeasts generally produce less than lager yeasts. Different strains of lager yeasts can produce as little as 3 ppm or as much as 16 ppm SO_2 in beer (Dr. Ludwig Narziss, *personal communication*, August 2004).

b. Microbiological contamination. The presence of spoilage organisms can promote high levels of sulfur compounds.

c. Copper vessels. Sacrificial copper from copper vessels or piping can catalyze the reduction of sulfur compounds.

d. Wort quality.   Wort deficiency of FAN, vitamins, or other micronutrients can lead to elevated sulfur levels. However, excessive levels of these compounds (including zinc) might also promote development of sulfur compounds.

e. Trub carryover.   High levels of trub in cold wort can promote H_2S production. However, if wort is deficient in lipid materials, SO_2 levels may be elevated.

f. Wort aeration.   Adequate wort aeration promotes healthy yeast growth and lower volatile sulfur compounds. Inadequate aeration will increase sulfur production.

g. Fermentation health.   Strong yeast growth will result in a lower final level of sulfur.

h. Fermentation vessel design.   Tall unitanks yield lower levels of sulfur compounds because of the greater scrubbing action of convection currents.

i. Fermentation temperature.   Cold fermentation may increase sulfur levels.

j. Cellar additives.   Any materials added to beer should be examined for the presence of sulfur compounds. Some materials such as isinglass may have been stabilized with SO_2. Although SO_2 is a powerful antioxidant, it can be reduced by yeast to form the rotten egg-smelling compound H_2S.

k. Scrubbing action of CO_2.   Some brewers strip volatile sulfur compounds from their beers after primary fermentation by bubbling carbon dioxide gas through the aging tank. Other brewers pump the fermented beer into the top of a "drop receiver" vessel and out through the bottom while transferring to the lager cellar. This causes vigorous splashing, which flashes off some volatile compounds.

l. Yeast autolysis.   An increase in H_2S levels late in fermentation may be due to yeast autolysis.

m. Culture age.   Freshly propagated, healthy yeast produces very little SO_2. After the first generation, however, the yeast cells generally produce their "strain typical" levels of SO_2 (Dr. Ludwig Narziss, *personal communication,* August 2004).

n. Oxidation.   Air pickup during beer handling lowers volatile sulfur levels.

84. What is the significance of acetaldehyde in fermentation?

Acetaldehyde is the aldehyde found in the highest concentration in beer and has a "green apple" character. Its taste threshold is 10 ppm, and in finished beer, it is found typically at 2.5–20 ppm. Most aldehydes originate in the wort (from malt) or through beer staling. Acetaldehyde, on the other hand, is an intermediary product of fermentation. Early in fermentation, it is excreted by the yeast; it is later reabsorbed during fermentation and aging. It has the characteristic flavor of green apples or grape skins and—along with sulfur compounds and VDK—is associated with immature beer.

85. What factors influence acetaldehyde production in fermentation?

a. Yeast strain.   Yeasts vary in their ability to produce or reduce acetaldehyde.

b. Wort FAN levels.   If wort lacks sufficient amino acids, acetaldehyde levels may increase.

c. Microbiological contamination.   Contamination with bacteria, especially wort spoilers and *Zymomonas*, can increase acetaldehyde levels. The gram-negative *Zymomonas* bacterium can be a significant problem in primed, cask-conditioned ales, due to its vigorous production not only of acetaldehyde but of hydrogen sulfide, as well.

d. Fermentation temperature.   High fermentation temperature promotes acetaldehyde production. However, this can be readily reduced during warm aging.

e. Overpitching.   Too high a pitch rate can promote acetaldehyde production.

f. Fermentation health.   Any factor that favors a healthy fermentation will aid the reduction of acetaldehyde in late fermentation and aging.

g. Insufficient aging.   All factors that promote strong secondary fermentation during aging can reduce acetaldehyde levels.

Table 1.5 shows the production factors that may affect formation of certain volatile by-products.

86. What is attenuation?

Attenuation means "thinning down." During fermentation, the transformation of extract causes a thinning of the liquid and a lowering of its specific gravity. Attenuation is expressed as "apparent" or "real."

Apparent attenuation is influenced not only by the lowered extract, but also by the formation of alcohol. The density of water at 20°C is 0.998

Table 1.5. Control factors for beer volatiles

	Esters	Higher alcohols	Volatile sulfur
High wort gravity	+ +	+	
Low wort aeration	+ +	−	+
High wort aeration	−	+ +	−
Repeated aeration	−	+ + +	−
High pitch rate	−	−	
High fermentation temperature	+ + +	+ + +	
Very low fermentation temperature	+	−	+
Yeast strain selection	± ± ±	± ± ±	± ± ±

+ = Slight increase. + + = Moderate increase. + + + = Strong increase. − = Slight decrease. ±, ± ±, ± ± ± = Increase or decrease, depending on the strain.

g/ml, while that of ethanol is only 0.789 g/ml. The formation of alcohol, therefore, lowers the apparent gravity.

87. Why is attenuation important?

Brewers track apparent attenuation during the fermentation process because it is an indication of the speed of fermentation. Brewers want fermentation to proceed in a consistent and predictable manner. Specific gravities should be measured and logged at least once per day during main fermentation. This information allows the brewer not only to monitor the speed of fermentation, but also to determine when to adjust tank temperature, "bung," "fass," or harvest yeast. Knowing the original wort gravity and final attenuation allows the brewer to calculate the final beer alcohol level.

88. How is attenuation measured?

Attenuation is measured with a hydrometer, refractometer, or densitometer. Refer to Chapter 5, Volume 2, for details on use of these instruments.

89. What is the real degree of attenuation"?

The real degree of attenuation, also known as the real degree of fermentation (RDF), is the percentage of the total wort extract that has actually been fermented. Alcohol must be removed (for example, by distillation), followed by topping up to the original volume with distilled water before measuring real extract. See ASBC *Methods of Analysis,* Beer-5.

90. What is the apparent degree of attenuation?

The apparent degree of attenuation, also known as the apparent degree of fermentation (ADF), is the percentage of the total wort extract that is "apparently" fermented. Because of the presence of alcohol, the wort's apparent degree of attenuation appears higher than the real degree of attenuation. ADF levels typically fall between 75 and 85%. Low ADF levels result in beers with a raw, "flour-like" flavor. High-ADF beers may still have a full flavor because elevated ethanol levels somewhat compensate for the lower dextrin concentration.

91. What is the attenuation limit?

The attenuation limit is the gravity of beer after all of the available extract is consumed. Brewers should routinely compare their beer's final apparent attenuation with its attenuation limit in order to ensure that fermentation is complete. In general, beer bottled for the trade should contain minimal residual fermentable extract in order to ensure biological stability. The residual fermentable extract in beer brewed for bottle refermentation also must be known before the sugar or wort additions can be calculated.

92. How can brewers measure the attenuation limit of a beer?

A simple rapid fermentation can be performed by collecting 200+ ml of aerated wort from the plant wort cooler. Wash and filter healthy yeast slurry through a Buchner funnel or centrifuge the slurry sample at 1,000 rpm for 5 minutes, discarding the supernatant. Weigh out 15 g of the thick yeast. Mix the yeast and wort in either a 500-ml flask or 250-ml fermentation tube ("Gärrohr," available from VLB Berlin Laboratory equipment catalogue). These containers should be loosely covered to discourage evaporation. Accelerated results can be achieved utilizing a wrist shaker or a stirring bar on a bench-top mixer (*Figure 1.12*). A few drops of antifoam can be added if needed. Ferment for 24–48 hours at 68°F (20°C). If more yeast is used (16 g/100 ml of wort), results can be obtained in about 7 hours. After the yeast has settled, determine apparent gravity. The fermentation tube works well for these methods because, once fermentation is complete, the yeast settles tightly into the base of the tube (*Figure 1.13*). See also ASBC *Methods of Analysis*, Wort-5.

Figure 1.12. A platform shaker can be used for either wort forced-attenuation measurement or for laboratory propagation of brewer's yeast. (Photo by Cynthia Stalker)

93. How can the freezing point of beer be estimated?

If the alcohol by weight (A) and the real extract (RE) are known, the freezing point (in degrees Celsius) can be estimated as follows (Hardwick, 1994, p. 655):

$$\text{Freezing point } (°C) = 0°C - (0.42\,A + 0.04\,\text{RE} + 0.2)$$

Yeast Propagation, Handling, and Pitching

94. Why should brewers propagate their own yeast?

Without healthy yeast, proper beer cannot be made. The availability of numerous commercially produced yeast strains in pitchable quantities (along with turnkey brewing systems) has been the driving force behind the success of the pub brewing industry. However, for small specialty breweries that package, maintaining healthy yeast requires a greater investment. Historically, breweries that purchase yeast slurries from outside

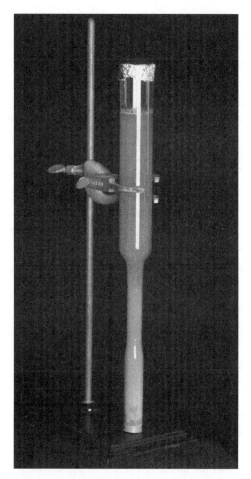

Figure 1.13. Fermentation tube designed for simple, rapid determination of wort final attenuation. Knowing wort final attenuation is critical for traditional lager brewing. (Photo by Cynthia Stalker)

sources often "inherit the sins" of the previous owner. If a specialty brewer is committed to long-term success, there is no reason why yeast needs cannot be met with in-house propagation.

95. How often should yeast be propagated?

In order to minimize the effects of contamination and yeast mutation, most lager brewers repropagate a fresh culture after three to 10 fermentations, or "generations."

Some brewers repropagate for every batch. Modern cellar practices cause added stress to yeast, and using fresh, vigorous yeast allows for predictable fermentations. Breweries using unique top-cropping strains such as Bavarian weiss-bier cultures in tall, closed vessels might also find benefit in continuous propagation.

Other brewers (particularly top-fermentation brewers) may have been continually repitching the same yeast culture batch-to-batch for hundreds or possibly thousands of generations with no apparent change in fermentation performance. The decision on how often to introduce a new propagation into the plant depends on the strain (ale yeast is typically more robust), fermentation performance, yeast-handling practice, flavor and aroma characteristics, changes in flocculation, attenuation, plant design, bacterial or wild yeast contamination, and the brewer's philosophy.

96. How is pure yeast propagated?

The first pure yeast propagation system was commissioned in 1885 by Professor Emil Hansen in conjunction with the Carlsberg Brewery. There are many variations on this original design, but the principles remain unchanged. Propagation is divided into three steps: the culture maintenance phase, the laboratory phase, and the plant phase.

Examples of typical propagation procedures are as follows. When working with culture yeast, absolutely sterile techniques must be used at all times.

a. Culture maintenance phase. Propagation requires a source of pure culture yeast, which can be supplied by a yeast bank. Normally, the proper strain, often supplied on nutrient agar slants, is shipped from the yeast bank by air freight. This "mother culture" is used to inoculate wort agar slants to be used as the "working culture" for in-house propagation. The working slants should be stored in a designated refrigerator at 34–39°F (1–4°C). Because yeast cells lose viability and can mutate over time, it is best to prepare only a four- to six-month supply.

Many smaller brewers do not maintain the mother culture in-house but instead purchase fresh slants from a yeast bank for each propagation cycle. It is possible, however, to maintain the mother culture—or even numerous working culture vials, purchased from a university or similar institution. In this case, the cultures can be stored on growth media containing 10–20% sterile glycerol and frozen at or below –112°F (–80°C).

b. Laboratory phase. It is a good practice to streak a wort agar plate from a stock slant culture and then select only healthy-looking

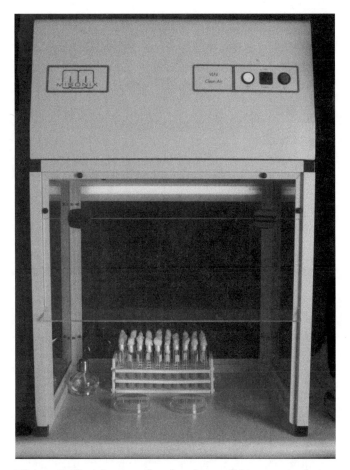

Figure 1.14. A laminar flow hood is useful in any size brewery. It can be used for either laboratory yeast propagation or microbiological testing. Note the presence of working yeast culture tubes under the hood. (Photo by Cynthia Stalker)

colonies to work with in the propagation. A loop full of yeast from a healthy colony is combined with approximately 50 ml of sterile hopped wort at 5–7°P and 68°F (20°C). *Figure 1.14* shows a laminar flow hood for microbiological work in the lab. This wort is fermented for 24–48 hours to at least slightly past the peak cell growth phase to ensure maximum cell count and acclimation to all fermentable wort sugar types. The culture is then transferred to 500 ml of 12°P wort, and the same growth pattern is repeated. Multiple flasks can be run in parallel. Serial step-up dilutions of 10–100× are performed every 24–48 hours until there is suffi-

Figure 1.15. The Carlsberg flask typically has a capacity of 20 L. In this vessel, laboratory wort is sterilized, cooled, aerated, and inoculated with culture yeast. After propagation, the flask is taken into the plant, and the culture is inoculated into a fermenter or yeast propagator. (Courtesy of Alfa Laval Inc.)

cient volume for the plant phase of propagation—for example, to 20 L in a Carlsberg flask (*Figure 1.15*). Any greater growth in the laboratory would become cumbersome. *Figure 1.16* shows a schematic of the laboratory phase.

In order to be successful, a proper yeast laboratory separate from the production area should be designated. If such a space is not available in the brewery, the laboratory propagation can be performed, for example, at the microbiology department of a local university.

c. Plant phase.   The laboratory culture is inoculated into either a well-sanitized production fermenter (if no propagation plant exists, see *Figure 1.17*) or into a designated yeast propagation system (*Figure 1.18*).

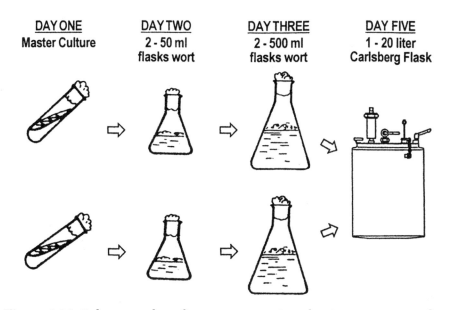

Figure 1.16. Laboratory phase for yeast propagation, showing process steps from slant to Carlsberg flask.

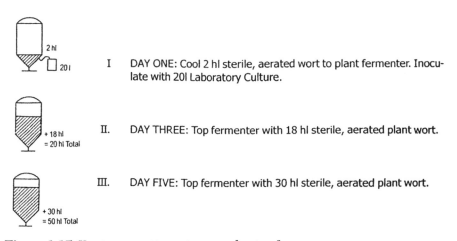

Figure 1.17. Yeast propagation using a production fermenter.

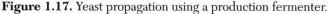

The culture is topped with sterile aerated wort every 24–48 hours, increasing the volume five to 10 times for lager yeast and 10 to 25 times for ale yeast until sufficient yeast is produced for a normal fermentation. As the propagator is topped, the temperature is slowly dropped to the necessary plant fermentation temperature. Most brewers do not wait until the yeast flocculates but pitch the propagator contents soon after the fer-

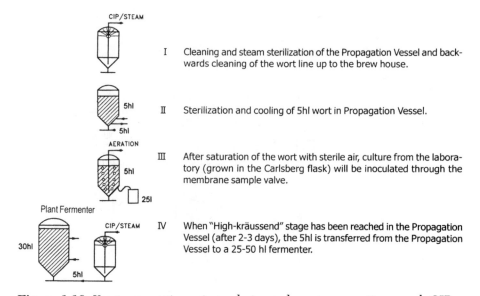

Figure 1.18. Yeast propagation using a designated yeast propagation vessel. CIP = clean-in-place. (Courtesy of Alfa-Laval Inc.)

menter peak cell count is achieved. Using a modern propagation system with heavy aeration, it is possible to produce healthy yeast at 80–200 million cells per milliliter, which can be pitched into as much as 40 times that amount of wort. A typical propagator size for a small specialty brewery might be 0.5–5 bbl. The total time from "yeast-slant-to-production fermenter" can take as little as 4–7 days.

97. What types of propagation plants are available to the specialty brewer?

Modern propagation plants are designed for maximum cleanability and sterility. Small specialty breweries use single-vessel, two-vessel, or continuous proportional-fed batch systems.

The single-vessel system (*Figures 1.18* and *1.19*) utilizes one vessel to sterilize, aerate, and cool wort, followed by inoculation with the lab culture in the same vessel. Laboratory repropagation is therefore required every time yeast is propagated.

The two-vessel system, on the other hand, utilizes one vessel for wort sterilization/cooling and a second for actual propagation. The brewer can leave a portion of the fresh culture behind in the propagation vessel after pitching to restart the process again without having to resort to laboratory step-ups.

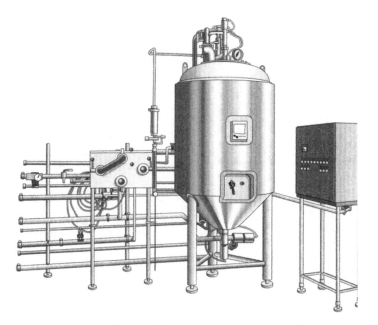

Figure 1.19. Single-vessel yeast propagator; note the top agitator plate, piping, and flow panels. These systems can sterilize wort prior to propagation pitch to ensure propagation purity. (Courtesy of Alfa-Laval Inc.)

The two-vessel system is more expensive to purchase and is better suited for continuous propagation with one yeast strain (*Figure 1.20*). The single-vessel system is sufficient for most specialty breweries because continuous propagation is not normally necessary and because many specialty brewers use multiple strains of yeast. Small brewing plants generally propagate fresh yeast only weekly, monthly, or even less frequently. *Figure 1.21* shows a yeast propagation plant installed at a large specialty brewery.

A popular system for the smaller brewers is a 65-L, two-tank propagation plant designed by David Bryant (*Figures 1.22* and *1.23*). This design is based on a continuous proportional-fed batch system (CPF system). The first vessel is used for sterilization of the wort and the second for the actual propagation. Laboratory culture (500 ml) is inoculated into the propagation vessel. Using a simple peristaltic pump and programmable logic controller (PLC), an ever-increasing flow of wort is dosed from the sterilization vessel to the propagator over 2–5 days with continuous aeration. The feed of sterile wort is controlled so that it is proportional to the yeast's growth rate in an exponential fashion. This system

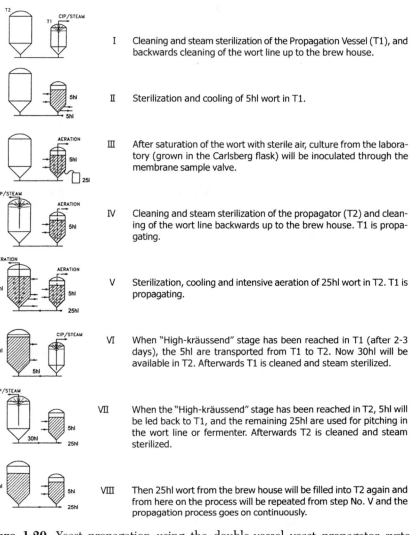

I   Cleaning and steam sterilization of the Propagation Vessel (T1), and backwards cleaning of the wort line up to the brew house.

II   Sterilization and cooling of 5hl wort in T1.

III   After saturation of the wort with sterile air, culture from the laboratory (grown in the Carlsberg flask) will be inoculated through the membrane sample valve.

IV   Cleaning and steam sterilization of the propagator (T2) and cleaning of the wort line backwards up to the brew house. T1 is propagating.

V   Sterilization, cooling and intensive aeration of 25hl wort in T2. T1 is propagating.

VI   When "High-kräussend" stage has been reached in T1 (after 2-3 days), the 5hl are transported from T1 to T2. Now 30hl will be available in T2. Afterwards T1 is cleaned and steam sterilized.

VII   When the "High-kräussend" stage has been reached in T2, 5hl will be led back to T1, and the remaining 25hl are used for pitching in the wort line or fermenter. Afterwards T2 is cleaned and steam sterilized.

VIII   Then 25hl wort from the brew house will be filled into T2 again and from here on the process will be repeated from step No. V and the propagation process goes on continuously.

Figure 1.20. Yeast propagation using the double-vessel yeast propagator system. (Courtesy of Alfa-Laval Inc.)

produces vigorous yeast growth, yielding a cell count of more than 150 million cells per milliliter and thus making the system suitable for pitching up to 20 bbl. Total propagation time from "slant-to-fermenter pitch" can be as little as four days.

98. How is yeast harvested?

The type of process and method of yeast harvesting has a great influence on the consistency and quality of the yeast slurry. If handled

Figure 1.21. Yeast propagation tanks and equipment installed at a specialty brewery. (Courtesy of Sierra Nevada Brewing Company)

Figure 1.22. Continuous proportional-fed propagator. The wort sterilization vessel is on the right, and the yeast propagation vessel is to the left. This design is well suited to the needs of small brewers, especially if multiple yeast strains are used. (Photo by Cynthia Stalker)

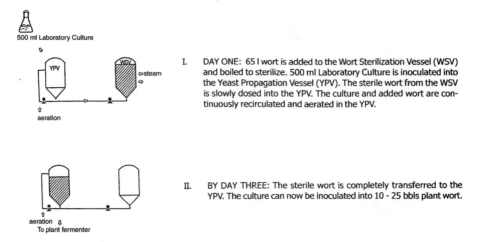

I.   DAY ONE: 65 l wort is added to the Wort Sterilization Vessel (WSV) and boiled to sterilize. 500 ml Laboratory Culture is inoculated into the Yeast Propagation Vessel (YPV). The sterile wort from the WSV is slowly dosed into the YPV. The culture and added wort are continuously recirculated and aerated in the YPV.

II.   BY DAY THREE: The sterile wort is completely transferred to the YPV. The culture can now be inoculated into 10 - 25 bbls plant wort.

Figure 1.23. Yeast propagation using the continuous proportional-fed system.

properly, top-cropped ale yeast can be relatively free from trub, have high viability, and have a consistent cell count since it is virtually all yeast with little entrained beer. With shallow, flat-bottomed fermentation vessels, the first head (formed in 12–24 hours) contains most of the coagulated protein and dead yeast cells. This is typically discarded, and the second crop that forms is harvested after 60–96 hours. Toward the end of fermentation, this yeast is typically almost pure yeast, with very high viability. With small fermenters, the yeast can be scooped manually off the surface and then placed into storage brinks. Alternately, the yeast head is allowed to collapse and be harvested off the bottom similar to the harvesting of lager yeast, although this yeast is typically not as clean and contains sedimented cold break and dead yeast cells. After primary fermentation, the beer would be transferred directly to casks (or, now more commonly, to closed tanks) for further maturation and conditioning.

Lager yeast is usually removed from shallow fermenters after the tank has been emptied of the beer, utilizing specially designed scrapers that may have small raised feet that allow selective recovery of the yeast sediment while leaving the earlier deposited trub behind. Some breweries may harvest ale fermenters in a similar fashion.

Conical tanks allow the removal of sedimented yeast and cold break to take place with the tank still fermenting and full of beer. Trub, cold break, and sedimented yeast should be removed as soon as practical after filling with wort and the start of fermentation (ideally, within 12 hours) to prevent the reabsorption of unwanted material. Practices vary as to when to collect the sedimented yeast for repitching, but most brewers discard early-flocculating cells since they have proven that they are not up to the task. One or more yeast discharges, or "cone blows," are typically performed before the clean yeast is allowed to sediment for ultimate collection. Cropping is usually timed close to the end of fermentation, when the yeast is at its healthiest. Care should be taken to determine the maximum rate to discharge from the cone to prevent blowing beer through the center of the sedimented yeast. After cropping, additional yeast continues to sediment. It is a good practice to regularly purge this, since some tanks may have trouble maintaining cold temperatures in the bottom of the cone, and autolysis of the yeast increases the pH, degrades beer quality, and provides fuel for bacterial growth.

The brewer should never leave flocculated yeast in the bottom of a fermenter in the mistaken belief that the yeast will aid in further maturation. Once the yeast has settled out of solution in the primary fermenter, it is of little benefit to the beer.

99. What equipment is utilized for yeast storage?

Storage vessels, or yeast brinks, can be as simple as a clean bucket stored in a cooler or refrigerator or as elaborate as an automated tank and agitation system. Rapidly cooling the yeast after harvesting is crucial in any storage system to help maintain viability. A modern, closed yeast brink typically contains a sanitary, built-in, low-shear agitator; cooling jackets and temperature control; sanitary sampling device; load cells for measuring contents; and built-in CIP. Newer installations may use double-seat tank outlet valves to allow for hygienic cleaning and flushing between pitch cycles. Tanks should ideally have shallow cone bottoms, although dished or flat pitched designs are also in use. Some brinks may also have sterile air or, preferably, inert gas for positive top pressure and the ability to aerate the yeast with oxygen or air before pitching. Some designs incorporate gentle external pump loops instead of agitators for mixing. Small-diameter, high-speed agitators are not recommended; newer designs may incorporate slow-speed, multilevel large paddles contoured to the bottom. All systems should be optimized to do as much homogenization and as little damage as possible. Since all that is required is to achieve good mixing and cooling, systems can be programmed to cycle on or off as required to minimize shear stress on the yeast. During storage, air pickup should be minimized since it accelerates the depletion of the cell's stored energy.

Some yeast brinks also serve the dual purpose of being acid washing vessels, but after such treatment, the entire contents must be used, so a smaller dedicated tank may be preferable.

When harvesting tall fermenters, entrained CO_2 in the yeast causes excessive foaming in an atmospheric brink. Harvesting into a pressurized brink at the same pressure as the fermenter and then cooling, agitating, and degassing over several hours works well.

100. How long can harvested yeast be stored?

After fermentation, the yeast survives on stored energy reserves and fairly rapidly starts to lose vitality. Yeast is best stored at 32–39°F (0–4°C). Colder temperatures can prolong storage time, but care should be taken to prevent freezing. Storage time should be as short as practical; ideally, less than three days. Some breweries with limited production or after holidays or maintenance shutdowns may have to pitch yeast that has been stored for 5–7 days or longer, although with somewhat lower vitality.

101. Is it beneficial to aerate yeast prior to pitching?

Some brewers feel it is beneficial to aerate or oxygenate the yeast directly prior to pitching. Yeast brinks are sometimes equipped with an aeration lance and agitator, or some breweries transfer the stored yeast into a dedicated vessel designed to saturate and mix the slurry with oxygen or air several hours before pitching. This same vessel can also be used for acid washing, if required. This technique minimizes the lag time for the start of active fermentation, but once aerated, the yeast must be put into production within several hours to prevent the depletion of the cell's glycogen reserves, which are utilized for energy until the yeast can assimilate nutrients from the wort. Some breweries use this method in lieu of wort aeration, or with reduced wort aeration, with the belief that it minimizes oxidation of wort constituents. Another method some brewers use to activate the yeast prior to the main fermentation is to add approximately a 50% volume of well-aerated wort to the brink from 2–24 hours prior to the addition to the main fermentation.

102. What is acid washing?

Acid washing is the term used to describe the practice of acidifying or treating the yeast in an attempt to purify the slurry prior to repitching to minimize bacterial contamination. The underlying principle is to subject the slurry to conditions that will kill or inhibit bacteria but leave the yeast—which has the ability to tolerate lower pH levels than many bacteria—largely unaffected. For many years, brewers have debated the merits and demerits of this practice. Some feel it is used as a crutch for poor sanitation and hygiene practices and, since it is not effective on all strains of bacteria, it may do more damage to the yeast than benefit. Other breweries practice yeast washing every generation, claiming no detrimental effect on the culture, and some even claim improved fermentation performance after treatment.

Several different acids, times, and pH levels are in use, but probably the most common treatment is to acidify the chilled slurry with food-grade phosphoric acid, lowering the pH to 2.2 while constantly stirring for 1–2 hours at or below 39°F (4°C) directly prior to pitching. Some brewers add ammonium persulfate (0.75%, w/v) to the acid, which is reported to enhance the effectiveness. Unhealthy yeast should be discarded, not washed. Temperature and pH should be closely checked during this process, as yeast can be damaged if precise conditions are not met. Yeast should not be stored after washing.

Other yeast purification methods such as the use of chlorine dioxide have been advocated for use in place of food-grade acid for yeast washing. Although this or other methods may or may not work, care should be exercised whenever using products registered or approved for one use in the brewery but not for a different application, because of the potential for formation of harmful by-products or introduction of excessive levels of potentially harmful products.

Ideally, in a well-maintained and clean brewery, acid washing or other methods used to combat bacterial populations in the pitching yeast will not be necessary, although in some older facilities, brewers may feel they have few options. Good fermentation and yeast-handling practices encourage rapid and healthy yeast growth that inhibits bacterial growth and can go a long way toward maintaining insignificant levels of contamination in the culture yeast.

103. What are typical sources of microbiological contamination in the fermentation cellar?

a. Cropped yeast.   Bacteria and wild yeast collect in the yeast and can be carried on through serial repitching.

b. Wort aeration system.   Wort can easily back up into the wort aeration piping, check valve, flowmeter, and regulator.

c. Worn gasket materials. Gasket materials on tanks, piping, and valve seats are subject to wear due to contact with harsh cleaners or simply due to age. All gaskets in the fermenting cellar should be replaced at least once per year. When designing a cellar, it is best to minimize the use of gasketed flanges and hoses as much as possible. A welded "hard pipe" installation is more expensive up front but will yield dividends through superior performance and longevity.

d. Improper cleaning. Biofilms and beer stone build up on any beer contact surface (especially if that surface is scratched or rough). Proper CIP or COP (clean-out-of-place) procedures are required to prevent buildup in equipment. Special portable ultraviolet lamps can be used to help the brewer find beer stone. When UV light is shone onto a stainless steel surface (with no background white light), the beer stone fluoresces and can be easily identified. Such lamps are also used in the dairy industry to find milk stone.

e. CO_2 distribution piping. Gas distribution systems are prone to contamination, especially if beer should back up into the piping system. A proper CO_2 distribution system should be built from stainless steel, include sanitary sample ports, and be cleanable.

f. Deadlegs. Piping, pumps, and tanks should be built with cleanability in mind. Zwickels, relief valves, threaded fittings, extended tees, vertical pipe runs, check valves, nonsanitary pressure gauges, hoses, pipe reducers, gate valves, ball valves, globe valves, carbonation and aeration stones, and the like should be avoided or installed in such a fashion that they can be properly removed for hand cleaning.

g. Flaws in pipes, pumps, or vessels. Cracks or crevices in machinery harbor contaminating microbes.

h. Direct contamination. Malt dust, insects, or surface mold can be the vector for direct contamination of beer. Milling and brewhouse areas should be walled off and separate from cellars. Windows and doors should be screened against insect and rodent ingress. Moldy walls and floors should be cleaned and sanitized with a "residual"-type sanitizer. Keeping the cellar's relative humidity below 80% discourages mold growth. Moldy beer hose should be discarded.

i. Floors. Floors contain the highest concentration of beer-spoiling organisms in the brewery. All flooring should be designed for easy cleaning. Raw concrete is too porous to use in a cellar and should be covered with an epoxy or tile system. Cracks in tile or epoxy should be repaired.

104. Are any other treatments used to prepare the yeast for reuse?

Depending on how the wort was handled in the brewhouse and the method of yeast collection, the harvested slurry may contain varying amounts of dead yeast, coagulated protein, cold break, hop resins, and other undesirable particles. Historically, brewers would wash the harvested yeast with cold sterile water and sometimes screen out impurities over a fine vibrating mesh screen. Due to the increased risk of contamination and to improvements in wort production, this practice is now not as common as it once was, with greater attention being paid to hot and cold wort clarification and harvesting methods that minimize inclusion of these undesirable materials. There has been some renewed interest in treating yeast with a vibrating screen or other gentle agitation methods to help degas the yeast in breweries using tall fermenters. The resulting high hydrostatic head entrains excessive CO_2 levels in the yeast cell, causing stress that may inhibit subsequent fermentations; processing the yeast can help in the release of excess CO_2. Modern enclosed vibrating screens that include the capability for automated CIP are available. They are expensive but are a good method for removal of CO_2 and for cleaning yeast, which can lead to improved yeast vitality.

105. How much yeast slurry should be pitched?

Before determining the quantity of slurry to be pitched, the cell count, yeast viability, and trub content should be assessed. Some brewers forego any analysis and achieve acceptable results simply by pitching a known quantity or weight per barrel; they then make empirical adjustments based on the perceived thickness, prior fermentation history, and condition of the slurry. This typically ranges from 1 to 3 pounds of thick yeast slurry per barrel. This practice is still probably common in some small breweries, and as long as the fermentation rate, yeast performance, and characteristics of the beer can be maintained, this may be an acceptable method. Although not having consistent pitching quantities can lead to flavor and fermentation changes, even the smallest breweries can develop methods to help minimize this variability.

Several straightforward (although slightly tedious) procedures have been in use in the brewing industry for many years to determine cell counts in the harvested slurry. Most brewers have relied on the use of a graded slide or hemocytometer and a microscope to count individual cells in diluted yeast slurry. Once the cell count is determined, the desired pitch rate can be calculated. Care must be taken in the dilution steps to ensure homogeneity of the sample, or large errors can be introduced. This method also relies on the accuracy of the person doing the counting. Utilizing an inexpensive handheld counter and averaging several samples helps improve accuracy.

The development of automatic cell-counting equipment has allowed for much higher throughput and repeatability, although variability in dilution and sample homogeneity can still introduce significant errors. These systems are programmed to count particles only in a specific size range and may not as accurately count cells that are in clumps or budding.

Another simple method is to centrifuge a sample and determine the percentage of solids. This procedure can be used without any additional information to help standardize the pitch amount based on weight or volume. A sample of the yeast to be pitched is spun down in a laboratory centrifuge and, by using a graduated sample tube or weighing the clarified beer, the percent solids is determined. Since approximately 1 pound of yeast solids per barrel will give about 10 million cells per milliliter, a ratio of the percent solids in the sample can be multiplied to determine correct amount to pitch. Since cell size and desired pitching rates can be quite variable, this should be used with the addition of an occasional comparative hemocytometer calculation, so an actual cell count of the slurry can be approximated.

Ideally, all of these methods should be coupled with viability and trub assessment to get a true representation of the viable cell count in the pitching slurry. In order to correct for the presence of trub, 1 ml of a 50% caustic solution per 50 ml of yeast slurry can be added to the centrifuge tube. The caustic dissolves the trub, leaving only the settled yeast after centrifugation.

106. How is the viability of the pitching yeast determined?

The most widely used and rapid method involves microscopic examination, utilizing dyes that are absorbed into the yeast cell. Living cells break down the dye into a colorless compound, but the dead or dying cells retain the dye. Although still widely used, this methodology can be somewhat inaccurate, giving rise to either high or low false viability, particularly with weak or stressed cells. It is also possible that some cells may be damaged by the dye itself, leading to false low viabilities. Some degree of interpretation is required on the part of the technician since some cells may be only weakly dyed. Historically, the dye methylene blue has been used, but due to poor reproducibility with some yeast strains and stressed or low-viability samples, other dyes and techniques are used as well.

Also, a sample of the culture can be plated and the viable cells counted, although this test is probably not rapid enough for regular production needs and, for the small brewer, is probably necessary only for troubleshooting a unique problem or process.

New, rapid methods are being developed that employ fluorescent dying of the yeast culture. In this method, the dead cells show coloration under a specially equipped fluorescent microscope or sample cell. With further development, these methods may lead to simple, accurate viability assessment.

Laboratory direct-reading capacitance meters have recently been introduced that promise to give a direct reading of the viable cells in a yeast slurry without any dilution and minimal lab manipulation.

107. What is yeast vitality?

Vitality refers to the health and fermentation ability of the yeast cell. Most methods used to determine whether a cell is alive, or viable, cannot readily assess the condition of the yeast or the level of stress to which it may have been subjected. Cells that have undergone long or poor storage conditions or damaging physical or environmental conditions can suffer a loss of vigor, and although alive, may not produce a normal fermentation.

Monitoring the pH of the yeast slurry is a rapid method that gives an indication of the overall condition of the yeast; as the yeast degenerates, the pH rises. Some brewers set an upper limit of 0.5 pH units above the fermenter pH level; yeast reaching this limit is not used for repitching.

Tests can be performed to assess the vigor of the culture. One is based on the acidification ability of the yeast during a pilot fermentation, using the rate and pH drop to assess the overall vitality of the culture. Another simple method is to measure the pressure increase under standard conditions in a closed vessel during a defined time period (Dr. Ludwig Narziss, *personal communication*, August 2004). Advanced staining methods and other laboratory analyses can also be used.

108. How are pitching and fermentation temperatures determined?

The type of beer being produced and the overall fermentation and aging cycle, as well as the philosophy of the brewery, dictate the pitching temperature. As some brewers work to maximize throughput and decrease fermentation time, increased pitching temperatures and changes in methods of fermentation and aging have evolved.

Pitching temperatures are typically several degrees cooler than fermentation temperature. Since fermentation is an exothermic process, the pitching temperature is selected to allow for a natural "free-rise" to the desired fermentation temperature and then controlled to produce the desired flavor compounds, which are produced (and in some cases also reduced) in varying amounts depending on the temperature regime and yeast strain. The pitching and fermentation temperatures are arrived at by balancing the pitch rate, pitch temperature, maximum fermentation temperature, fermentation cycle, fermenter design, and, most importantly, the desired flavor profile. Some amount of experimentation with many of these variables will probably be required to establish the optimum combination.

109. What requirements must be met to help maintain healthy yeast and proper fermentation?

a. Proper pitching rate for the selected strain.   A pitching rate that is too high prevents adequate new cell growth. A rate that is too low may encourage bacterial growth.

b. Correct fermentation temperature range for the selected culture yeast.   Ale strains should be fermented at 60–72°F (16–22°C). Lager strains should be fermented at 40–55°F (4–13°C).

c. Sufficient initial wort aeration.   Most strains need an O_2 level of at least 8–10 ppm. The level of aeration required depends somewhat on the strain and on the original gravity.

d. Proper carbohydrate composition.   Adequate levels of fermentable material, primarily glucose, maltose, sucrose, and fructose, must be supplied.

e. Proper level of FAN in the wort.   If the FAN level is too low, it will limit yeast growth. If it is too high, it may encourage bacterial growth. It is typically not a problem to maintain a sufficient supply of amino acids for all-malt wort.

f. Adequate levels of Ca, Zn, Mg, and other trace elements

g. Prevention of oxygen pickup during yeast storage, reducing glycogen depletion.

h. Proper yeast harvesting and storage.   Yeast should be harvested and stored so as to minimize trub inclusion, storage time, and temperature.

i. A clean plant, to minimize contaminants

Cold Break in the Wort

110. What is cold break?

Hot wort out of the whirlpool is normally clear but becomes cloudy upon cooling. Cold break, or cold trub, is a loose, brown coagulant formed in wort once it is cooled below 158°F (70°C). The greatest concentration forms below 68°F (20°C). Cold break consists of about 50% protein and 25% polyphenols; the balance is carbohydrates and lipids. Unlike hot break, which forms in the brewhouse, cold break particles are extremely small (less than 1 micron in diameter)—about the size of bacteria.

111. Is cold break harmful to beer quality?

A small amount of cold break in fermenting beer is beneficial. Cold break is a source of yeast nutrients such as sterols and unsaturated fatty acids. Cold trub acts as points of nucleation for removal of dissolved carbon dioxide from the fermentation. Carbon dioxide is inhibitory to yeast, so some CO_2 removal, coupled with the subsequent convection currents in the fermenter, promote a more vigorous fermentation.

However, excessive levels of trub (particularly hot break) can decrease foam stability, decrease shelf stability, decrease beer filterability, and contribute a harsh flavor to beer. Trub can also coat yeast cells and interfere with their metabolism, which is a concern when using the modern practice of serial repitching without yeast screening or washing.

112. How can brewers measure cold trub levels?

Collect a sample of cold wort immediately after wort cooling but before yeast pitching. Add 100–250 ml to a glass graduated cylinder. Let the sample settle for 3 hr at fermentation temperature. Note the wort clarity as well as the amount and character of the sediment. The volume of the sediment is normally between 2 and 3% of total volume, assuming that the kettle boil, fining, and whirlpool operation are optimized. If no kettle fining is used, the cold wort might be cloudy (greater than 200 EBC haze units) and the volume of trub may approach 0%. This should not be considered problematic as long as results are consistent and the beer flavor is as expected. Results greater than 5% suggest that the hot break was not properly separated or that too much kettle fining was used.

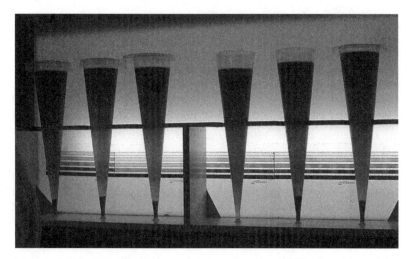

Figure 1.24. Imhoff tubes, used to measure the cold trub produced (in milliliters per liter). (Courtesy of Sierra Nevada Brewing Company)

A 1.000-ml Imhoff tube can also be used for this test (*Figure 1.24*), as its design allows for excellent sedimentation and therefore very accurate determination of the sediment volume. Gently rotating the Imhoff tube during sedimentation aids this process. However it is difficult to characterize wort clarity as the sample thickness varies with depth in the Imhoff tube (Ian Ward, *personal communication,* May 2005).

In the brewing literature, cold trub levels are often cited in milligrams per liter, not milliliters per liter. In order to determine wort cold trub levels in milligrams per liter, refer to MEBAK *Brautechnische Analysenmethoden,* Band II, 2.7.

113. Should brewers attempt to remove cold trub from wort?

In the past, German lager brewers placed great emphasis on control of wort cold trub levels in order to ensure smooth flavor and superior beer clarity. A separate cold trub removal step does not appear to be part of the Bohemian brewing tradition. In modern "streamlined" breweries, cold trub removal is not normally practiced. The use of highly modified malts, green beer centrifuges, and silica gel–polyvinylpolypyrrolidone (PVPP) stabilization reduces to some extent the negative effects of cold trub. Many brewers believe that cold break removal has very little, if any, impact on modern beer quality.

There may be a benefit, however, for the brewer to remove some cold trub if brewing pale lagers. The extra steps required may yield subtle improvements in mellowness of flavor and haze stability. Each brewer is encouraged to study the effect of cold trub removal in his or her own brewery to determine whether there is a benefit. A simple test procedure would be to fill a unitank with cooled, aerated wort but delay yeast pitching for 8–16 hours. The settled, cold trub is then carefully removed from the tank cone, and yeast is injected from, for example, a pressurized keg. If correct sanitation procedures are followed, there should be no problem with wort spoilage. The finished beer flavor can be compared to the standard (Dickel et al., 2002; Kessler, 2002).

The brewer should be cautioned that hot trub carryover from the whirlpool is a significant concern, and its control is paramount in quality brewing.

114. If desired, how can brewers control cold trub levels?

Removal of cold break is very difficult, because of its small size and "gummy" consistency. It is neither practical nor desirable to remove 100% of the cold trub, but there are at least six methods of removing some of this trub. Because cold wort is a rich source of nutrients for spoilage bacteria, any equipment used for such a process must be of sanitary design.

a. Skimming the foam layer from open fermenters.   Some of the cold break will rise to the surface on the foam layer. This first heading, or *brandhefe,* is composed of dead yeast and protein. It can be skimmed off open fermenters. Closed fermenters can be designed and filled to ensure overfoaming to allow for the deposition of *brandhefe* on the head of the tank or for overflow from the tank vent.

b. Use of cone-bottomed fermenter.   Cold break can be purged from the cone of a cylindroconical fermenter along with dead yeast during the first 8–12 hours of fermentation. The convection currents of active fermentation after 8–12 hours make trub removal difficult, although it is still possible to remove dead yeast.

c. Filtration with diatomaceous earth.   Filtration removes 75–90% of cold break when performed at 32°F. This process is rarely used, because of its inherent complexity and because fine, gummy, cold trub easily plugs the filter. However, filtration allows for the control of wort trub levels by the use of partial filter bypass.

d. Centrifugation.   Centrifugation removes 30–70% of cold break. As with filtration, the wort trub concentration can be adjusted by par-

Figure 1.25. Flotation tank in a small German brewery. A whirlpool vessel can also be seen on the left and a small clean-in-place tank to the right.

tial bypass. The centrifuge, however, is rarely used to remove cold trub, because it is difficult to separate the fine cold trub particles from the heavy wort.

e. Starter tank. A starter tank is normally a shallow (1 meter deep), flat-bottomed vessel. The cold wort is simply held in this shallow vessel to allow the cold trub to settle. If the starter tank wort contains yeast, it is not held for more that 12 hours because, once fermentation commences, the movement of the wort does not allow for further sedimentation. After 12 hours, the wort is transferred to the fermentation tank, and the settled trub and dead yeast are discarded. About 35% of the trub can be removed. If no yeast is added to the wort in the starter tank, the wort can be held for as long as 16 hours to achieve 50% cold trub reduction. The wort is then aerated and pitched while pumping to the fermenter. The long settling time is required because of the fine particle size of cold trub. However, deep, cone-bottomed starter tanks are sometimes used in modern automated breweries, and the wort is held for as little as 1 hour.

f. Flotation tank. Because the use of a traditional starter tank requires 12–16 hours and a large footprint, the flotation process was borrowed from other industries to achieve better results in less time and in a smaller space. A flotation tank is a simple cylindrical tank that can operate properly, even if over 12 ft tall, thus saving floor space (*Figure 1.25*). In the flotation process, the cold wort is aggressively aerated with a specially designed compressed-air dosing system (e.g., a "two-compartment jet" aeration device) to create numerous fine air bubbles. A flow rate of 20–60 liters of air per hectoliter is required, and pure oxygen should never be used. Any excess undissolved gas slowly rises to the wort surface, carrying along with it 50–65% of the coagulated cold break into a thick foam layer. After a rest of 2–6 hours, the wort is pumped to the fermenter, leaving behind the top flocculated foam-trub layer. This process has the added benefit of allowing for a second wort aeration, if desired, as the wort is pumped to the fermenter.

Cellar Design and Construction

115. What are the basic building requirements of a fermenting cellar?

Depending on the type of beer to be produced and the philosophy of the brewery, the design of fermentation cellars can take many different forms. All cellars need to be constructed with materials that allow for easy cleaning and maintenance and that can withstand the rigors of constant water, detergents, sanitizers, and steam. Mold and bacteria thrive in the humid environment found in many cellars, so construction materials must withstand the required constant cleaning and harsh environment.

a. Drainage. Floors should be adequately sloped to drains, ideally with a slope of at least ¼ inch per foot (2 cm/m). Many contractors try to save on the cost and trouble caused by multiple slopes and drain connections, but this is not a wise place to try to save money. Well-designed drain locations and slopes will do much to enhance the cleanliness and hygiene in the brewery. Even a very good concrete or flooring contractor will have problems ensuring that there are no "bird baths" with slopes much less than ¼ inch per foot. Tearing out and redoing a poorly drained floor is much more costly than doing it properly.

If using individual drains, the floor is typically laid out in grids with one drain serving each "square." These grids are typically limited to 20–25 ft from the drain to minimize the distance that solids like yeast must travel and also to limit the depth of the drain from the edge of the top of the slab. When calculating slope, remember to use the distance from the hypotenuse or corner of the square for slope calculations, particularly if the slope is less than ¼ inch per foot. Trench drains can allow for a simpler floor layout and also make for easier hosing and cleanup. However, they require greater maintenance and cleaning than individual drains and should be designed with easily removed grates to facilitate regular removal and scrubbing. Pre-sloped trench drains are available in stainless steel and several composite materials and can be ordered in sections to allow for different slope and sewer pipe configurations.

The drain section must be well encased in concrete to ensure structural integrity of the floor. Alternatively, Styrofoam forms can be purchased with a pre-engineered slope and cast into the floor giving a seamless drain system that is then coated with epoxy. Individual floor drains should be equipped with drop-in baskets to minimize the problems caused by glass, gaskets, or other foreign material that may clog the system. Well-planned cleanout locations should be installed to allow for snaking out the system when required.

Domestic waste piping should be kept separate from the production waste drainpipe to minimize potential problems in case of a blockage and to facilitate the ability to pretreat production waste without having to deal with the potential problems caused by having to treat potentially pathogenic domestic waste.

Waste piping materials must be able to withstand the high temperature and corrosive nature of beer, brewery wastewater, and co-products. Lengthy 180°F (85°C) sanitizing rinses, caustic and acid CIP solutions, and even beer can damage some materials used in conventional waste systems. Some local municipalities may have requirements calling for cast-iron waste piping for commercial construction. Modern thermoplastic materials are much easier to install, and in most cases, they can handle higher corrosion and temperature requirements. In areas that may be subjected to strong oxidizers such as concentrated peracetic acid or hydrogen peroxide, standard acrylonitrile butadiene styrene (ABS) may not be suitable. More-suitable polyvinylidene fluoride (PVDF), polypropylene, or other specialized drain, waste, and vent material should be specified.

In CIP rooms, the use of stainless steel materials may be warranted until the waste stream can be buffered with other lower-temperature flows. Boiler blow-down water should be tempered before being discharged into the waste system. Piping should be adequately sized and sloped (¼ inch per foot) to prevent flooding of areas when high-flow-rate activities such as tank rinsing or cleaning take place. The cost difference is fairly negligible, for example, between a 4-inch and a 3-inch pipe system.

Once under the slab, retrofitting is very difficult and costly. The design and requirements for venting and trapping a floor drain system can have some unique requirements. Not all plumbers and building inspectors are familiar with some of the allowable methods, which can create unnecessary problems during construction. Select a contractor who has knowledge in this area and proven past experience.

b. Floor finish.   Because of the constant abuse and the near impossibility of optimum repair conditions, floors are one of the more difficult areas to maintain in the brewery. Heat, chemicals, beer, and impact damage from hoses and other fittings take their toll on almost all flooring materials. The most proven modern materials for brewery flooring are tile and seamless troweled epoxy or similar synthetic materials. Tile is typically more expensive than troweled coatings and may not be as durable or chemical-resistant as modern synthetic materials. The multitude of grout joints can be harder to keep clean than a seamless floor. Tile has some advantage in that small repairs can sometimes be made without being noticeable or requiring extensive downtime. Epoxy grout and setting material can be used with tile instead of cement-type or asphalt-type grouts to increase durability. Nonslip tile finishes are available with Carborundum or silica grit embedded in the surface, although these coatings may eventually wear down and lose their effectiveness. Quarry tile has typically been used in breweries and is available as bricklike pavers, which typically are more than 1 inch thick and can withstand extremely heavy traffic, or as 5/8-inch-thick material, which is suitable for areas with moderately high traffic. Quarry tile can be damaged by the concentrated acids and oxidizers commonly used at breweries; porcelain or other high-fired tile materials may be more resistant to these chemicals and are available in a wide range of colors and sizes, but they are much thinner and usually not intended for heavy traffic. Troweled, seamless synthetic finishes are available in a variety of base materials. They have varying levels of ability to withstand thermal shock and impact and varying levels of chemical resistance. Consult with a reputable coatings contractor to ensure that the material is correct for the application.

Finding an experienced flooring contractor is probably the biggest factor in the long-term success of the flooring system. A skilled applicator can resolve minor slope problems and create coves on equipment pads or other sanitation enhancements. Troweled flooring will eventually need to be recoated and repaired. The floor needs to be dry and usually sanded to allow good adhesion of a new finish coat, and this is sometimes problematic in an operating brewery. Dust and chemical fumes may require that the area be cleared of all vulnerable product and personnel for several days. Grit is usually rolled into the finish coat to provide a nonslip surface; the entire floor area may need to be refinished to color match. Floor preparation is the key to a durable floor with any system. Blast tracking with steel shot is the preferred preparation method, although acid washing fresh, non-oily concrete is acceptable if done correctly. Moisture from fresh concrete or underlying moist soil conditions can make adhesion of any flooring material fail. A suitable vapor barrier or crushed-rock vapor break is recommended under the slab. Taping a 2-ft square of Visquin down on the bare concrete floor for several days may help reveal potential moisture problems. If, when the plastic is removed, visible moisture is observed, a surface-applied elastomeric membrane or a flooring material that is specifically designed for this condition may be required. Both tile and troweled floors are available in a wide range of colors and can be inlaid to add interest.

Rolled-on epoxy paint is acceptable in low-traffic areas but must be recoated regularly, which can be difficult during production periods. Both water- and solvent-based materials are available, although air pollution limitations in many areas may limit choices to water-based products only. Varying durability with both systems has been reported, but concrete preparation again is a key factor to the success of the installation.

Bare concrete is not acceptable in most production areas, since cleaning solutions and acids in beer readily attack it. Surface-hardening additives can be added to the concrete before pouring or can be sprinkled on during finishing, helping to improve chemical and physical integrity in areas subjected to occasional exposure.

All floor surfaces must be easy to clean but have an adequate nonslip finish to prevent injuries from slips and falls. Some breweries have used diamond plate stainless steel for flooring in areas that are prone to damage or have high sanitation requirements. Although the initial installation cost may be significantly higher than for other options, the long-term savings and lack of production interruptions may warrant its use in areas such as under a CIP skid or in a clean room for a bottle filler.

c. Walls.   Ideally, walls should be finished with materials that are easy to clean and impervious to impact, water, and chemicals. Glazed tile over concrete, plaster, or Sheetrock makes an attractive and very durable combination, although much more expensive than other acceptable options. Some food plants opt to leave concrete walls and ceilings bare in an attempt to minimize the ongoing maintenance and possible downtime requirements of a painted surface. If properly prepped, however, epoxy paint will last many years and will prevent the possibility of mold and bacteria from becoming established in the porous concrete surface. Paints are available that contain mold inhibitors and may help control growth in hard-to-access areas. Ideally, walls should be constructed of solid, non-rotting materials, such as solid concrete, concrete block, brick, or pre-glazed block.

Common construction methods utilizing wood or metal studs with Sheetrock or plywood sheeting are less expensive to construct, but they will invariably be problematic because of the near impossibility of eliminating all moisture ingress from around wall and floor joints. At a minimum, this type of wall should be constructed on raised 6-inch coved concrete curbs, and covered with resilient products such as smooth fiberglass-reinforced plastic (FRP), which works much better than painted Sheetrock and is less costly than tile. If Sheetrock is used, it should be "green board," which is somewhat stronger and more water-resistant than standard Sheetrock and only slightly more expensive. Hard-wall, smooth, concrete backer board is much more durable than Sheetrock and makes for a very acceptable finish when smoothly troweled with plaster and painted with epoxy. Polyurethane, silicone, or other flexible caulking should be used between floor and wall joints and should be regularly inspected and maintained. Some caulking products are also available with mold inhibitors that help maintain hard-to-access-and-clean areas.

d. Ceiling.   Depending on the building design, ceilings can either have a hard surface (such as concrete or steel decking), with lighting and piping systems suspended below, or a suspended ceiling below the pipework, ducting, and conduits. Both designs have merit, and one may be the better solution depending on the overall plant construction methods and cellar system. Pipes, beams, and other overhead objects will trap dust and, with some construction materials, may be difficult if not impossible to keep clean. Ideally, the building materials should be able to withstand a thorough wash-down on a regular basis. With buildings constructed utilizing concrete ceiling decking, a suitable finish may be achieved without any further surface finishes other than painting, or in some cases, the brewery will opt for bare concrete. Concrete poured over "B" deck galvanized steel can also suffice as a surface finish as-is, or it can be painted.

However, this is not the most attractive or sanitary finish, and the galvanized coating can be attacked by CIP vapors.

All roof systems eventually leak, so care must be taken to maintain or proactively repair the roof, particularly over open production areas, to prevent contamination. Some breweries are constructed utilizing metal building columns and roofing, with just a vinyl-coated insulation blanket over the perlins (ceiling joists). This, although not ideal, can suffice in areas with closed fermentation vessels. Occasional vacuuming of the inaccessible area above the perlins is about the only thing that can be done to help minimize dust buildup, since this type of roof system cannot tolerate frequent water wash-downs. This ceiling design is undesirable over open fermenters or other open-process areas.

Suspended ceilings have the advantage of providing a flat, smooth surface that minimizes problems caused by dust and other overhead contamination caused by pipe sweating and condensation, although dust buildup above the ceiling area (or between cracks or loose tiles in some designs) can cause a problem during maintenance. Ceilings can either be suspended on a grid system or be structural, allowing access above for piping and maintenance. Grids can be purchased in corrosion-resistant materials such as anodized aluminum or plastic. By utilizing FRP panels and clips or silicon adhesive, ceilings can be made dust-tight. Prelaminated FRP plywood can be attached to the underside of steel or wooden trusses to give a smooth structural ceiling finish.

e. Equipment.   Tanks, pumps, and pipe supports are best handled by building up a small, 2- to 6-inch-high housekeeping pad that is slightly sloped and designed not to impede floor drainage, with the tile or epoxy flooring material coving up the base of the pad. When possible, equipment can be supported several inches above the floor on stainless pipes directly embedded or core-bored into the slab before the final floor finish is applied. This minimizes the problems that can be caused by water being trapped behind a larger housekeeping pad and makes for a clean and less expensive installation. In some areas, earthquake requirements and local building officials may dictate the mounting and anchoring methods.

f. Lighting.   Lighting should be bright to assist in operations and help in visually inspecting and maintaining the work space. All fixtures in production areas should be designed for wet environments and include gasketed covers and wash-down capabilities. Relatively inexpensive covered fiberglass fluorescent fixtures are available in several sizes and wattages. High-pressure sodium lights with vapor-proof fixtures and gasketed lenses are very energy efficient, although the color spectrum is not ideal to work under. High output 250- to 400-watt waterproof and vapor-proof "white light" metal halide fixtures are available that offer a more pleasant working environment, but these may take several minutes to warm up. (However, quick restrike kits are available.) Approximately 100 foot-candles will give good working illumination. In some production areas, natural lighting supplied via skylights provides a superior (and inexpensive) source of lighting, as well.

g. Electrical service.   Outlets must be equipped with waterproof covers, and some jurisdictions may require ground-fault interruption. Any motors should be designed for wet environments, with switches, disconnects, and electrical boxes designed to NEMA (National Electric Manufacturers Association) 4X or 12 standards, or even to IP67 standards (suitable for total submersion). Since voltages are typically 240–480 three-phase, electrocution or shock hazards are great, so every precaution must be taken to minimize exposure. The typical wet cellar environment is extremely hard on most electrical equipment; it is worthwhile to spend extra on specifying stainless steel or corrosion-resistant, coated conduit since galvanized materials are attacked by the caustic detergents commonly used in breweries. Epoxy-coated, "dirty-duty," or stainless steel motors can also be purchased that have additional gasketing and seals offering more protection than standard TEFC (totally enclosed fan cooled) designations.

116. What type of ventilation is required for a fermentation cellar?

Carbon dioxide buildup can be fatal! Signs of carbon dioxide suffocation start with dizziness, fatigue, and then unconsciousness. Relatively inexpensive, continuous CO_2 monitoring and alarming units are now available and should be incorporated into the cellar design; these can also be used to trigger exhaust fans to automatically eliminate high levels of CO_2. Small personal monitors are also available but should be used in conjunction with room-mounted units for greater safety. With open fermentation systems or cellar designs that vent some or all of the fermentation gas into the room, an adequate clean or sterile air supply and CO_2 removal system are required. Exhaust ducts located close to floor level are useful in removing the heavier CO_2, with excess air supplied at ceiling level. Ideally, with open fermenters, the room should be supplied with positive pressure-filtered and sterilized air. Systems can be purchased or built that incorporate cooling coils and several stages of filtration that can achieve greater than 95% removal of airborne contaminants. This can be followed up with an ultraviolet light chamber effectively delivering sterile air.

With closed cellar tanks that are vented through a CO_2 foam trap to the atmosphere or connected to a CO_2 recovery system, air change requirements may be less severe. Cellar designs utilizing uninsulated or unrefrigerated tanks must rely on room cooling to maintain tank and cellar temperatures. Glycol or direct-expansion refrigeration coils can be located in the cellar, or dehumidified chilled air can be supplied from an external source. To help reduce mold growth, additional humidity control may also be necessary. Cellar designs with totally enclosed, individually jacketed and cooled tanks may need only moderate air changes, and conditioning the air may be required only for operator comfort.

Cellar Tanks and Equipment

117. How does fermenter design affect beer flavor?

Since the cost per barrel is significantly less with larger tank sizes, the trend has been toward larger tank volumes, both in conventional and modern fermentation systems. Transportation difficulties in moving large-diameter tanks and in fabricating large-dished heads to increase volume have driven tank designs taller as the volume demands increased. There are also limitations to providing adequate cooling in tanks of very large diameter, since the surface-to-volume area may be inadequate, and the distance from the center of the tank contents to the edge is great. Brewers have found flavor and fermentation differences with taller, large-tank designs, requiring the manipulation of fermentation parameters to try to duplicate more-traditional tank designs. Large tanks can develop significant convection currents, producing accelerated "stirred" fermentations that produce rapid gas evolution. This may result in scrubbing of both undesirable and desirable volatile compounds. Tall tanks also subject the yeast to high hydrostatic pressures, which affect the yeast's metabolism and performance characteristics. Large tanks may also create undetected temperature stratification, with areas that are much warmer than others, leading to fermentation and flavor changes. Since many brewers are very cautious about changing any production variables, some new cellars are designed with relatively small tanks. These breweries—after gaining practical experience dealing with problems with very large, tall fermenters—have opted to install smaller tanks in future expansions.

118. How does pressure affect fermentation?

The taller the fermentation tank, the higher the hydrostatic pressure in the bottom of the fermenter. Some brewers think that depths of more than 10–15 ft (3–4.5 m) may have an effect on fermentation. Static head is developed, 1 psi for every 2.3 ft of beer level in the tank. Brewers incorporating natural conditioning may also operate with an additional top pressure of 8–20 pounds during the conditioning process. Elevated pressures can inhibit yeast growth and tend to restrict the formation of some higher aliphatic alcohols and esters. Aromatic alcohols such as 2-phenylethanol are not strongly affected. Convection currents caused during fermentation and cooling will most likely equalize dissolved CO_2 levels, but after prolonged storage, the beer at the bottom of the tank may exhibit higher carbonation readings. Since testing is usually performed on beer taken from sample valves located at the bottom or lower part of the tank, the sample beer may not be a true representation of the tank contents.

119. How is the temperature of fermenting beer regulated?

The fermentation process is exothermic, generating about 250 BTU per pound of fermented extract. In very small tanks, air conditioning of the fermentation cellar can sometimes control this heat generation, but, more typically, the tank has some means for heat removal, utilizing coils or jackets or occasionally an external cooling loop. The amount of jacket area provided and the location of the jacketing is the source of some debate and much variation, depending on tank geometry, size, and manufacturer. Many models have been developed to try to simulate the ideal cooling jacket locations and circulation currents during active fermentation and cooling cycles. Due to the complexity of these models and the variability of fermentation parameters caused by different beer types, tank sizes, and conditions, most jackets are significantly oversized since correcting an undersized jacket after installation is not practical. A typical range for a 100- to 400-bbl tank with an intermediary coolant such as propylene glycol is 0.5–1 ft²/bbl, although great variation exists. Some brewers may utilize different combinations of jackets during different stages of fermentation and aging to promote convection currents or to provide cooling of the yeast sediment in the cone during the later stages of fermentation. The size of the cooling jacket is normally dictated by the rate at which crash cooling is specified at the end of the fermentation process, since this is the heaviest refrigeration demand.

120. What are the different types of cooling jackets?

Historically, tanks that were not in refrigerated cellars utilized internal cooling coils with chilled water or brine to control fermentation temperature. For improved cleaning and sanitation, most new tanks utilize external cooling coils or jackets that are welded to the outside tank wall or shell. Tank manufacturers may prefer one design over another—or tout the benefits of their proprietary manufacturing process—but the key to good cooling performance is adequate surface area, flow volume, proper velocity, and location. Three common jacket designs are typically available. Channel jackets can be continuously formed into half pipes or rectangular shapes and rolled and welded onto tank shells, creating a spiral jacket; however, applying this design to a small-diameter cone section can difficult. Preformed dimpled plate jackets can be formed and plug-welded to the tank wall. Internal baffling or separate zones may be required for large heat-transfer areas to ensure proper velocities over the entire jacket. Automatic resistance-welded, inflated jacketing is produced by at least one manufacturer. This design has some limitations as far as jacket volume and cannot be installed on thicker tank walls, but it can be a less expensive alternative to other jacket options for some applications.

121. What are the common types of refrigerants for fermentation cooling?

Jackets can be designed to utilize either a secondary refrigerant such as food-grade propylene glycol or a direct-expansion refrigerant such as Freon or ammonia. Direct-expansion refrigerants can give higher electrical efficiencies due to lower evaporator pressures and compressor horsepower. However, the large amount of refrigerant that is required to circulate in tank jackets causes environmental or safety issues in case of a leak and is sometimes avoided in new installations. Systems utilizing flooded-plate evaporators, with propylene glycol as a secondary heat-transfer medium, can achieve almost comparable efficiencies. The suction temperature of the compressor stays within a few degrees of the desired glycol temperature, which also allows the use of a very small volume of refrigerant. Secondary-refrigerant temperatures need to be maintained at close to (but not much below) the freezing point of beer, 28–30°F (–2 to –1°C) to prevent the possible localized freezing of beer during long periods of cold storage.

122. Where should cooling jackets be located on fermentation tanks?

Cooling jackets should be positioned so that the surface is covered when the normal fill volume is in the fermenter. Large rectangular tanks may require much of the sidewall area to be covered. Since the volume-to-surface area can be great, the bottoms are rarely refrigerated due to the insulating value and effect on yeast sediment. Smaller tanks should have the cooling area evenly distributed on the sidewalls.

Conical fermenters typically have a jacket or jackets located directly above the sidewall-to-cone seam and, depending on tank size and height, require one or more upper zones with independent temperature control. Cone cooling is usually required to control fermentation temperature and, more importantly, for temperature control if the tank is used for yeast storage or crash cooling. During crash cooling or as the beer temperature approaches its greatest density at about 36°F (2.2°C), the contents of the tank may "flip," with the warmer beer sinking to the bottom. Without cone and lower sidewall cooling, it may not be possible to achieve cold-lagering or cold stabilization temperatures. Differential temperature set points on the jacket zones may be required to promote cooling by encouraging the circulation caused by the differing densities. The cone jacket should be located as close to the bottom as practical, to help ensure that the thick, insulating layer of yeast sediment in the cone is kept cold. On small-diameter, steep cones, this can be challenging for some tank fabricators and jacket types, but it is an important consideration in the overall tank design. *Figure 1.26* shows the location of the jacket on an uninsulated tank.

123. How is beer drawn or racked from the fermenter?

Flat-bottom fermenters are usually designed to leave yeast sediment and trub behind, by use of a stand pipe on the tank outlet. Conical fermenters require that the yeast be removed before the tank is emptied of beer. In some designs of dished-head or conical-bottom tanks, an adjustable "racking arm" is provided to allow for drawing the beer off above the yeast sediment. This hard-to-clean extra fitting may be unnecessary with a

Figure 1.26. Location of the dimpled jacket on an uninsulated tank. Note the placement around the side manway door. (Courtesy of Sierra Nevada Brewing Company)

properly designed conical bottom, adequate yeast storage, and proper handling.

124. What process-piping systems are used to transfer beer in the cellars?

All hoses and piping used for beer transfers should be sized to minimize turbulence and pressure drop (ideally less than 5 ft/s) or approximately 35 gallons per minute (gpm) for a 1½-inch line and 50 gpm for a 2-inch line size. Whatever system is employed for transferring beer in the plant, procedures should be established to ensure that all product-contact surfaces are sterilized before use and cleaned properly afterward. If proper procedures are developed and followed, any process system can produce a top-quality product. Loops should be established before filling or emptying, to allow the recirculation of hot water or sanitizer through the entire beer path for sufficient contact time to ensure adequate sterility. In situations that do not have double-seat valve protection, hot water (180°F, or 85°C) is recommended to minimize any chance of product contamination.

a. Hoses. The simplest, least-expensive, and possibly most-flexible method of filling and transferring beer utilizes movable hoses that can be used for filling, yeast harvesting, emptying, and CIP cleaning. Several styles and materials are available for "brewer's hose." Some designs are mandrel-wrapped and have the advantage that hose ends can be incorporated during the construction and essentially bonded to the hose, although the liner is not as smooth or durable as designs incorporating seamless one-piece liners. Extruded seamless-liner hoses require a barbed fitting to be clamped on, using either hose-type clamps or (a more secure and sanitary option) a fitting that is internally or externally swaged onto the hose. Some swaged hose ends are designed to be at least partially or totally reusable, but these are still very costly. Swaged ends offer enhanced security for hoses that are used for supply pressure during CIP duty.

Hose material should be FDA-approved and free from rubber and other odors. New hoses may require several CIP cycles to be odor free before they are put into product service. Although most modern butyl rubber compounds used by manufactures have little odor, "pickling" of new hoses by circulating waste beer or yeast through them is a good safeguard to help ensure that any rubber odor is eliminated before they are put into service. Hoses used on the suction side of pumps should be designed to withstand sufficient vacuum, especially during hot sanitizing operations. To withstand collapsing or kinking, hoses typically contain reinforcing ribs or ridged plastic wire. Steel wire reinforcing is not recommended since it can be permanently deformed if it is kinked while hot.

Good-quality hoses can last several years but do have a limited life and should be inspected on a regular basis, as internally split or blistered hoses can be a significant source of contamination. Removable hose ends can trap beer and be a source of hidden contamination, and the exposed cloth from the wrapped construction of most hoses will absorb beer. Since the ends of the hose are subjected to the most stress, the ends can be removed and the hose shortened if still in good shape. After use and cleaning, hoses should be stored off the ground and either capped, stored in sanitizer troughs, or allowed to free drain.

Figure 1.27. Pipe fences and swing panels for beer, yeast, and CIP in a specialty brewery. (Courtesy of Sierra Nevada Brewing Company)

b. Pipe fences.   Although transferring beer through hoses is still common in many breweries, the majority of new installations utilize fixed hard pipes, or fences (*Figure 1.27*), because of the enhanced sanitation, labor reduction, and security they afford. Various designs of piping fences are used; the simplest designs are just a horizontal pipe(s) running in front of the tank valve with a valve and "tee" connection and valve(s) to block the flow in the fence. Sometimes the legs or other parts of the tank can be used to support the fence. A jumper connection is designed to allow the tank to be connected to multiple pipes or ports on the fence to allow for filling, emptying, CIP, and yeast cropping. If all the pipe fittings are positioned in an arc, one fixed-length pipe can be used for all connections. Alternatively, an adjustable two-piece swing bend can service all the ports. A small bleed valve can be installed on the jumper close to the tank outlet valve, allowing hot water to flow almost up to the tank outlet during hot sanitizing. This design is somewhat prone to alignment problems, particularly with long runs, since the piping expands and contracts significantly during hot sanitizing and cold-beer processing. Fence designs using

Figure 1.28. Valve matrix block for automatic sanitary valves. (Courtesy of Sierra Nevada Brewing Company)

fixed panels with accurately located ports solve most alignment problems and allow for easier installation of proximity sensors to give feedback signals for more-automated installations. Some fence designs may incorporate other services such as CO_2 supply, CO_2 collection, and air. Some companies provide prepiped sections that may simplify field installation. The choice of fittings, valve types, and location is a personal, philosophical, and process decision. Designs that locate the tank outlet valve at a remote location may cause problems in high temperature conditions, with the beer sitting in unrefrigerated and difficult-to-clean piping.

    c. Double-seat valve systems.   Fully or partially automated fermentation-cellar piping systems are used in some breweries, allowing tank processes to be automatically or remotely controlled. With the advent of hygienic, cross-contamination-secure, double-seat valves, many different processes can safely and simultaneously be conducted through a manifold of valves (*Figure 1.28*). Breweries have made significant investments in this technology, but, for small breweries, the high initial costs of equipment and associated installation and control are hard to rationalize. Larger breweries can better justify the benefit from the reduced labor costs and the ability to incorporate control logic (with possible production safeguards) into the process. Hybrid systems utilizing pipe fences or panels with automated double-seat tank outlet valves allow remote filling of fermenters, with the ability to sanitize and clean all the process piping up to the base of the fermenter. Even for the smaller brewer, bright beer tanks and transfer systems are good areas for this technology since tank residence times are short and hygiene requirements are demanding.

125. What is the best material for fermenters?

Historically, fermenters have been built from materials that could be manufactured with the equipment and technology of the time. Over the years, wood, stone, concrete, copper, glass-lined or epoxy-coated steel, aluminum, plastic, and stainless steel have all been used with varying degrees of success. All of these materials have some inherent problems either with fabrication, corrosion, cleaning, maintenance, or cost. Today, most brewery fermenters are fabricated out of stainless steel. Although stainless steel is initially more expensive than most other materials, its durability and its ease of cleaning and sanitizing make it the ideal choice for fermenter applications in the brewery. Lined or coated mild steel is still occasionally specified for large tanks. The applied surface coatings can be very smooth and easy to clean, but long-term maintenance issues generally outweigh the initial lower-cost benefits. Although not an ideal material, inexpensive polyethylene or polypropylene plastic is sometimes used to line older wooden tanks or to manufacture very small fermentation vessels. A skilled fabricator can heat-weld seams and fittings, producing a relatively sanitary surface, although plastics can be somewhat porous and easily damaged. Although many grades of stainless steel are quite robust and will give years of service if treated properly, they can be susceptible to corrosion or damage from the chemicals commonly used in breweries. Many sanitizers and chlorides found in water and other cleaning compounds can cause devastating damage to most common grades of stainless steel. Care must be taken in cleaning to monitor pH, chemical concentrations, and contact times and temperatures or to switch to more benign cleaning regimes to ensure the longevity of the tank (see Chapter 2, Volume 3, for a further discussion of tank cleaning).

126. What is the preferred grade of stainless steel for fermentation tanks?

Today, most fermenters are constructed out of type 304 stainless steel; it offers the best compromise between cost and the corrosion resistance required for a fermentation application. Type 316, which has improved corrosion-resistance properties, is sometimes specified, resulting in about a 25% increase in material cost. Types 2205, AL-6XN, and other exotic grades that have higher temperature- or corrosion-resistance are sometimes specified for more-demanding applications in the brewery. The use of these grades for safety and longevity is justified for some high-temperature applications, such as hot-liquor tanks and piping, but with proper care, type 304 should give a long service life for fermenter and other low-temperature and low-corrosion uses. Many insulation materials contain leachable chlorides and have been implicated in tank failures caused by stress-corrosion cracking. Moisture on the tank surface caused by the sweating on the tank wall from trapped moisture can wick and concentrate the chlorides from the insulation. Most problems have occurred in higher-temperature tank applications, but specifying surface-applied anticorrosion coatings, chloride-free styrene, or chloride-free Inswool (although of a lower insulation value than urethane) may be a wise safeguard.

127. What surface finish should be specified for fermentation tanks?

For stainless steel material and the complete fabricated tank, a variety of surface finishes and polish grades can be specified. As with most things in brewing, opinions vary as to what is best. The smoother the surface, the easier the tank will be to clean. With a smooth surface, fewer sites will be available where beer stone and bacteria can establish themselves. Thinner gauges (up to 10) leave the steel mill with a finish commonly referred to as 2B. This mill finish is a dull gray and not highly reflective. Under microscopic examination, the surface is usually seen to be fairly smooth, and when used in a tank, it is relatively easy to clean. Occasional surface irregularities and blemishes may be present from the mill, and, during fabrication and handling, great care is required to prevent surface scratches since any cosmetic or needed repair will never match the original finish. Welding of seams or fittings requires grinding and polishing, which may result in a patchwork appearance. For this reason, some industries specify a polished finish for the entire tank surface. Thicker material (or "plate") required for larger tank heads arrives from the mill with a rough finish and must be polished before use. Various polish grits can be specified, but typically 120- to 180-grit polish is suitable for most fermenter surfaces. These finishes are referred to by many trade names or specified by surface roughness or roughness average (RA) number; RA 32 is a common specification. Microscopic examination of these polished surfaces actually shows many scratch lines that some argue are more difficult to clean than the 2B mill finish. Higher polish specifications (RA 15), as well as electropolishing, are sometimes specified for yeast propagators.

128. How should new stainless tanks be prepared before they are put into service?

Most equipment manufacturers leave the final cleaning and tank preparation to the customer. During the fabrication process, the tank is contaminated with dirt, abrasive material from polishing, oil from machine tools, and mild steel pickup from fabrication equipment. All these contaminants must be properly removed before the tanks are placed into service. Most equipment manufacturers supply a recommended or required-for-warranty procedure to follow to clean and passivate their equipment before it can be used (most tank warranties have so many disclaimers you usually have little chance proving it was a material defect). The key steps are to first clean any soil, grease, or oil from the surface. After a preliminary visual inspection to ensure that all the protective paper and masking tape have been removed, a nonchlorinated caustic should be used at 2–3% and 140°F, or as recommended by the supplier. This removes oils and surface dirt. After rinsing, a thorough visual inspection should be made to ensure that all surface soil, dirt, etc., has been removed, since any oil left on the surface can be trapped in the passivated surface. The caustic treatment should then be followed up with an acid step to remove any free iron that may be on the surface and to speed up the formation of the thin "passive" oxide layer that makes stainless steel corrosion-resistant. Historically, hot nitric acid at 20–50% was used and is still probably the best treatment, although it is extremely dangerous to handle and difficult to dispose of. Many equipment and chemical suppliers now promote citric acid or a blend of phosphoric and nitric at much lower concentrations to avoid some of the dangers of hot nitric acid. Practical experience would suggest that they are not as effective as the hot nitric treatment, and citric acid may lead to off-flavor problems on the first fill. The resulting corrosion-resistant, passive surface can be removed by scratching, physical abrasion, or strong oxidizers such as chlorine, allowing the surface to be attacked and pitted. Much debate, misinformation, and conflicting theories surround the treatment of stainless steel surfaces. If clean and dry, stainless steel develops a natural passive layer in "air" after a matter of hours or days that is just as resistant as chemical passivation (see Chapter 2, Volume 3, for a discussion of passivation). Once passivated and given proper cleaning and sanitizing procedures, no other treatment should be required.

129. What is "beer pickling"?

To ensure that no off-flavors from new equipment are created in beer destined for the market, brewers sometimes "pickle" new tanks and equipment by brewing a batch of "waste beer" or circulating spent yeast or tank bottoms before the equipment is put into commercial service. This practice is not an exact science, and not all brewers have the same concerns. Even though chemical cleaning and passivation treatment are effective at removing most of the free iron and other contaminants from the surface, some brewers still detect flavor problems from new stainless steel equipment, especially equipment with a high surface-to-volume ratio, such as small tanks or kegs. It is now known that iron in very low concentrations can catalyze reactions leading to increased oxidation. Even after industry-accepted passivation, higher iron levels may be found in beer from fresh stainless steel equipment. Possibly other reactive ions in the fresh stainless steel may be a factor as well.

130. What fittings are found on fermenting tanks, aging tanks, and piping?

In the simplest designs, fermentation tanks may contain a minimal number of fittings; possibly just one fitting used for filling and draining the vessel. Any required testing would be performed by dipping out a sample to determine temperature, gravity, and possibly microbiological checks.

Modern fermenters typically contain numerous valves and fittings, making the life of a brewer much easier, as well as enhancing safety, sanitation, and control.

a. Tank inlet/outlet valves.   Several styles of valves are commonly used for filling and emptying fermenting vessels. Butterfly, diaphragm, plug, or double-seat valves are typically used for filling and emptying tanks. Valves can be located either directly on or close to the tank bottom or cone or remotely on the piping supplying the tank. Remote-mounted valves have the problem of trapping beer in piping that may not be temperature controlled or that may not have conditions identical to those of the fermenter. Double-seat tank outlet valves can allow the supply piping to be cleaned and sanitized up to the base of the tank. Butterfly or other single-seat valves can be more challenging to make sanitary between filling and emptying operations.

b. Sample valves.   Several different designs are available for taking beer samples. The most common design in use in the United States is the plug valve-style "zwickel," designed for use both for microbiological and physical analysis. Weld-in fittings with proprietary "bacteria-proof" gasketed seats or mini tri-clamp mounting are available. The valve must be "flamed" with a portable torch to render it sterile before microbiological tests can be performed. The only practical method to attempt to clean the valve after sampling, if the tank still contains product, is to hose or flush the valve externally. Flushing through the valve as part of the CIP regime is only marginally effective in cleaning the sample valve. Repeated heating of the valve causes the lubricant to degrade and caramelizes any entrained beer that is inherently trapped in the plug valve taper. The valve must be physically dismantled, soaked, and cleaned to ensure cleanliness. Lapping and lubricating the valve can help extend the service life but is a very labor-intensive procedure that is not regularly followed. This zwickel valve has recently been redesigned with a rubber diaphragm that can withstand reasonable levels of flame sterilization, as well as being easy and inexpensive to recondition. The main advantage of this valve design is the availability of industry-specific sampling equipment designed to directly mount onto the sample valve for physical testing, such as CO_2 and O_2 measurement.

Aseptic, diaphragm-type plunger sample valves can offer greater sample sterility, longevity, and ease of rebuilding, although it may be more difficult to obtain samples from them and, when they are used with some brewing-industry-specific sampling devices, proprietary adapters may be required. Several manufacturers produce similarly designed valves for use in the brewing industry; the basic design utilizes a rubber plunger that is submerged in a sterilant bath such as alcohol. Before sampling, the alcohol is drained and the valve can additionally be steam-sterilized if desired. After the sample is taken, the valve is flushed with water and packed with alcohol or other sanitizers until required for subsequent testing.

Conical tanks typically require that the sample valve be located above the cone section, or at least high enough to ensure that samples can be obtained above the yeast layer.

Several other options for sample valves are available, such as mini-diaphragm valves or valves using Teflon, rubber, or O-ring seals, but most have the same inherent problems with cleaning and sterilizing between sampling.

Septum-type valves for use with a syringe and needle are available for critical applications such as yeast propagation or critical microbiological work. These can be mounted in a special holder or on a standard tri-clamp fitting.

c. Pressure- and vacuum-relief valves.   Effective pressure- and vacuum-relief devices are critical for safe tank operation, since many tanks cannot tolerate any overpressure or vacuum without rupturing. CIP operations are particularly problematic since large pressure changes can occur almost instantly as the large gas volume inside an empty tank expands or contracts rapidly when rinses that are hotter or colder than the tank atmosphere are sprayed. Large tanks are often cleaned at less than 100°F or at ambient temperatures for this reason. Emptying and filling a tank without proper venting or counter-pressuring can also create excessive vacuum or pressure problems. Most jurisdictions in North America have adopted the pressure-vessel code requirements developed by the American Society of Mechanical Engineers (ASME). All tanks designed to operate above 1 atm (14.7 psi) or in vacuum service are generally required to be designed and certified (stamped) with an ASME approval. Depending on the local building department, county, state, or insurance carrier, stamped ASME pressure-relief valves may need to be fitted and inspected before the tanks are put into service. Many of the sanitary devices supplied to breweries may lack these certifications. If required by inspectors, a secondary relief device may be necessary. Any closed tanks must be designed and piped so that no possibility exists for accidentally creating excessive pressure unless the tank is suitably equipped with protective pressure-relieving devices. Like all fittings on tanks, relief valves and installation locations should allow for ease of maintenance and cleaning.

The simplest design for pressure- and vacuum-relief valves utilizes weighted devices that contain predetermined weighted seats or valves that will relieve pressure or vacuum when the tank pressure overcomes the equivalent static weight. These are simple devices, but they can become stuck closed and fail to open at the correct pressure if beer foam gets into the seat area and dries. Some of these devices contain air or CIP fluid-supplied operators or cylinders to lift the disk off the seat during normal tank cleaning. Some vacuum-relief valves typically found on "tank-

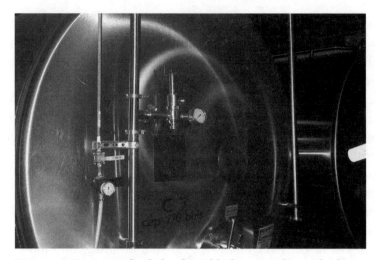

Figure 1.29. Spring-loaded, adjustable bunging device for keeping constant head pressure on a fermenter. (Courtesy of Sierra Nevada Brewing Company)

top plates" have an unfortunate design feature that creates a sump area. If the vacuum valve actuates, any liquid or material on the tank top may drain back into the fermenter, which can cause contamination if the tank contains product.

Other relief devices use springs against seats or a combination of springs and diaphragms to relieve excess pressure or vacuum. These may have wider latitude for opening and reseating than dead-weight devices. Systems that do not incorporate a seat-lifting device, and are out of direct CIP flow, should be inspected on a regular basis.

Many larger tanks are equipped with a large rupture disk that allows the immediate release of pressure or vacuum, since many mechanical relief devices may not react quickly enough to prevent implosion. If the tank has been specified with an adequate working reserve, a mechanical or dead-weight device may be selected to open at several pounds below the tank and rupture disk rating to allow the release of slight excess pressure without the expensive and more-disruptive immediate release of all the gas. Combination vacuum- and pressure-relief valves can be specified, and rupture disks are available that have an electrical contact that can alert the operator in case of rupture.

d. Bunging devices.   Some means of controlling fermentation or bunging pressure is required in most cellar operations. These can be either a spring-loaded type with an adjustable range (*Figure 1.29*) or a fixed dead-weight device. Bunging devices can either be located permanently on the pipework from each tank, and be designed for incorporation into the tank CIP, or be portable units that are moved from one tank to another. Some breweries may utilize a pressurized header(s) to both collect CO_2 and control back-pressure on fermentation tanks. Cellar tank pressure can also be controlled by utilizing a top-mounted pressure transducer with a process controller to modulate an analog-controlled gas valve or small butterfly valve to control tank pressure (see "Pressure gauges and sensors," below.)

e. Thermometer wells.   Most fermentation tanks contain one or more fixed probes to display or record tank temperatures. Welded-in thermo-wells allow for the insertion and removal of the probe when the tank is full of product or under pressure. Probe placement and length is dependent on tank geometry, size, and duty. Typically, probes are located away from cooling jackets, or with sufficient insertion length to minimize localized cooling effects. Thermo-wells should be long enough to provide a realistic temperature reading of the main tank contents. Common insertion lengths are 6–12 inches, but, for large tanks, insertion lengths of 18–24 inches may be required for accurate information. Some tank manufacturers weld the thermo-wells in at a slight vertical angle with the belief that it will help minimize the shadow caused during CIP cleaning. Several types of thermometers and sensors are available. A common size for both utilizes a thermo-well with a 0.265-inch bore and ½-inch female pipe thread, designed to accept a ¼-inch-diameter device. For local display, bi-metal thermometers are typically used. Three wire-resistance temperature devices (RTDs) are also used because of their high accuracy and reasonable cost. Small self-contained analog transmitters can be supplied in the head of the probe, providing a more friendly analog input for many PLCs. RTDs are also available with a local digital display. Since most probes have the sensing element at the end of the tip, wells with reduced-diameter ends give faster response time and still provide a robust installation. Utilizing a heat-conductive paste in the bore also helps increase response time. Like all sensors that are relied on for control or decision making, a periodic calibration program should be developed (see Chapter 1, Volume 3, for more discussion).

f. Pressure gauges and sensors.   Many tank piping designs incorporate fixed-pressure gauges or sensors to provide local or remote pressure readings. Standard open-tube Bourdon gauges are not recommended for most sanitary installations since product can become trapped in the coiled chamber. Many bunging devices contain integral gauges (some of these designs use open Bourdon tubes), or food-grade fluid-filled diaphragm gauges can be mounted on the tank vent piping. Many pressure gauges have built-in overpressure protection that is typically 150–200% of the pressure scale. If the pressure gauge is tied into the CIP supply of the tank, care should be taken in the piping design to avoid hydraulic pressure shocks ("hammer") caused by rapidly closing values or rapid startup of supply pumps. Supply pumps should be controlled by frequency drives or "soft start"-equipped controllers. Some tank cellars are equipped with pressure transducers to allow for the remote measurement and possible control of tank pressure. These diaphragm-strain gauge pressure transducers can offer accurate readings, with newer designs better able to handle overpressure without failure. With the addition of a lower-mounted pressure transducer(s), tank level and even density can be calculated and tracked.

g. Manway options.   All closed brewery tanks typically require an access hatch or manway—if not to allow for cleaning and inspection, at least to allow the welder to go home for a beer! Fermentation tanks designed for atmospheric operation may require only a gasket or seal capable of containing CIP fluid, but tanks designed for pressure service may require substantial design and engineering to restrain the extreme force.

Atmospheric manways are lightweight designs that are usually found above the liquid level but may be used in applications requiring low-head, water-tight sealing. They are typically of the out-swing style, utilizing either a flat gasket in the lid or a slotted donut gasket placed over the manway collar. Depending on the design, one or several lugs, wing nuts, or clamps hold the door closed.

Pressure-rated manways come in many types depending on the tank use and design. The most common pressure-rated manways are in-swing designs utilizing the tank's internal pressure to help maintain the seal. Better-designed doors have adjustable hinge points to allow the centering of the door and gasket. Some in-swing doors can be slightly challenging for an unskilled operator to maneuver. These designs typically have one central, large wing nut to hold the door sufficiently closed to maintain a seal. As the pressure builds internally, the sealing force increases. Excessive force applied to the wing nut to close a misaligned door can result in permanent damage. These types of manways can be designed to hold extremely high working pressures, but pressures over 1 atm (14.7

Figure 1.30. Cone manway access. (Courtesy of Sierra Nevada Brewing Company)

psi) require ASME certification. Usually, gaskets are either slotted or stretched over the edge of the door, or a flat gasket can be retained in a groove or depression. Insulated door designs are available to help with sweating on refrigerated tanks, but some versions have been known to develop internal cracks, possibly allowing contaminated beer to be trapped in the door and infect the tank contents. Frequent inspection or drilling of an external weep hole in the door is recommended for these designs. Since in-swing designs are generally not suitable for CIP, the manway is typically opened during the tank cleaning operation, with the gasket removed and cleaned separately. The design of some manways can create shadows under the opening, creating cleaning problems during CIP operations.

Several designs of pressure-rated out-swing designs are used in fermentation and aging tanks. The amount of hold-down lugs is dependent on the design and pressure rating. These are commonly used for horizontal chip tanks to help facilitate loading and removal of these materials.

On conical fermenters or unitanks, a removable swing-away cone section may be used in place of a manway (*Figure 1.30*), or the tank may have no lower manway and utilize a bolted tank access plate.

h. Tank-top plates.   Heavy, flat plates are usually bolted to a collar and counter-flange on the top head. Either a flat gasket or machined O-ring groove is used for sealing. Tank-top plates are sometimes used in place of manways. The underlying philosophy is that cleaning and sanitation are improved by eliminating the hard-to-clean lower manway that is submerged in the beer. The tank top contains all the required fittings and minimizes penetrations through the actual tank head. It typically contains a cleaning machine, pressure- and vacuum-relief valves, vent connection, sight glass, and inspection light. Some designs also incorporate a system for verification of cleaning machine rotation, since visual inspection is somewhat more difficult than with side manway designs. Sometimes a pressure transducer is furnished to monitor tank pressure and, when coupled with additional transducers, can also be used to track volume and density. These designs usually incorporate pressure and vacuum devices large enough to allow closed cleaning regimes, although larger tanks are usually cleaned at ambient temperatures or slightly above for safety. Periodic visual inspections should be scheduled that include the complete removal of the tank top, since the viewing windows in the tank top do not allow for an adequate cleaning assessment. Tanks without lower manways are difficult and potentially dangerous to enter. In addition to other confined-space requirements, proper training and the need for safety harness and evacuation winches are required. In very large or tall tanks, service personnel have resorted to inflating rubber boats and filling or emptying the tank with water to obtain access.

Tank-top plates incorporating sanitary agitators, aeration, and cleaning are also available to provide relatively complete and effective designs for yeast propagators.

131. How much headspace or freeboard is required in a fermentation vessel?

Ale fermenters typically have much greater freeboard requirements than lager beer fermenters because of the elevated temperatures and rapid gas evolution during fermentation. Wort composition, tank geometry, and yeast strain can also influence yeast head formation. Tall conical fermenters with high-gravity brewing may require up to 50% freeboard for ale production, but 25–35% is probably more typical. Lager fermenters may be designed with as little as 15% freeboard, but more typically, 25%. Many brewers try to achieve a slight overfoaming of the fermentation vessel in an attempt to deposit the undesirable *brandhefe* on the tank-top head in closed fermenters, since skimming the surface is usually impractical or impossible. To increase tank utilization, anti-foam additives are available, and various schemes such as spraying water or recirculation of beer through the spray ball have been tried, although most brewers are wary of these interventions. Controlling pitch rate and fermentation temperature can affect foam and yeast head development.

132. Why are traditional lager or aging tanks horizontal?

In the 1880s, Enzinger introduced the first practical filtration system for beer. Until that time, brewers prided themselves on achieving a relatively bright lager beer by a two- to three-month aging only. In fact, traditional brewers at that time did not immediately embrace filtration. They viewed it as likely to open the door to carelessness and hinder true progress. It was believed that a proper lager tank should be between 1 and 2 m deep. Anything taller than 2 m would not allow for complete sedimentation of yeast and colloidal materials. Tanks less than 1 m deep might cause premature settlement of the yeast. Therefore, relatively small horizontal tanks were the norm in lager brewing until progressive brewers determined that taller vessels could be used when combined with beer filtration. However, horizontal lager tanks are still very common in lager breweries today, and some brewers still see merit in the arguments of their predecessors of over 100 years ago. It may in fact be true that superior lagers are made in horizontal lager tanks, but the economic benefits of large, vertical vessels have made them the norm in modern breweries.

133. What is the advantage of a single-shell lager tank in a refrigerated cellar?

Single-walled tanks are cheaper to purchase than insulated, jacketed tanks. Single-shell lager tanks must be placed in refrigerated cellars. When oak lager tanks were used, the poor heat transfer from the cold cellar air through the thick oak staves meant that a freshly filled lager tank might take many days to fully attemperate to the ambient cellar temperature. This gentle cooling process allowed for a good start to secondary fermentation even if the lager cellar was cooled to freezing temperatures. However, when brewers replaced their oak vessels with aluminum or stainless steel tanks, it became difficult to properly control the lager temperature. If the cellar was too cold, the beer cooled too quickly and secondary fermentation was sluggish. If the cellar was too warm, the beer was not properly chill-stabilized. Therefore, single-shell lager tanks should be used only if, after maturation, the beer can be cooled to around 30°F before filtration.

Lager Beer Fermentation and Aging

134. How was lager beer made before artificial refrigeration?

The first artificial refrigeration machine was installed in a brewery in 1860. Before that time, lager brewing was restricted to the colder winter months. Often, lager cellars were built underground with a fermentation cellar directly above. Block ice was stacked on top or to the side of the fermenting cellar so that the cold air sank into the lager cellar, forcing warmer air out of the plant through duct work. The fermenting cellar was maintained at 42°F (6°C) and the lager cellar as close to freezing as possible.

Open wooden fermentation vessels had no attemperation coils and were small—about 30 bbl—to take advantage of the cold ambient temperatures. The beer was pitched at 1×10^6 cells per milliliter per degree Plato. The heat of fermentation developed at high kräusen caused the beer fermentation temperature to peak at about 46°F (8°C). After high kräusen, the beer temperature would slowly drop back to 42°F due to natural cooling from the cold cellar air. The majority of the yeast would flocculate and, if all went well, leave 2–10 million cells per milliliter and 1.0–1.2°P fermentable extract. Due to the cold temperatures and low pitch rate, fermentation lasted as long as 12 days. The beer was then dropped ("fassed") into a wooden lager tank that was left open to scrub out air and volatile sulfur compounds. (In German, *fass* means "barrel," and *fassen* is "to fill a barrel.") After 2–3 days, the tank was bunged. A second fermentation continued at a slower rate as the yeast consumed most of the remaining extract. The beer slowly cooled to near freezing. Over a period of 6–12 weeks, the beer matured, clarified, and naturally carbonated. The low fermentation temperature and pitch rate—coupled with long cold-lagering—ensured a full-flavored, smooth, estery beer. In the Bohemian tradition, not all of the fermentable extract was consumed—about 6% of the fermentable extract (0.7°P) was left behind, which contributed to the special character of those beers. *Figure 1.31* shows a cutaway drawing of a traditional ice-cooled barrelhouse.

Figure 1.31. A fermentation and lagering cellar (barrelhouse) from the late 1800s. Cooling was generated by block ice. Cooled air cascaded down into the cellars through ductwork. The brewer was able to control the temperature in the cellars as well as the evacuation of carbon dioxide simply by manipulation of gates in the ducts. Because such cellars were so dark and damp, lager brewers were quick to embrace artificial refrigeration once it became available.

135. How have lager brewing practices changed over the last 150 years?

Modern technology has allowed the lager brewer to produce clean, consistent beer in a shorter time due mainly to the following:

a. Artificial refrigeration means that consistent, stable beer can be produced all year. Stainless steel construction allows for large batch sizes—tanks with capacities of more than 6,000 bbl are possible.

b. Sanitary equipment and cellar design allow for warmer fermentation temperatures without fear of microbiological contamination.

c. Understanding of fermentation flavor development, especially di-acetyl, allows for greater control of fermentation parameters and shorter aging times.

d. Collection and availability of compressed carbon dioxide gas make adjustments to carbonation simple.

e. Beer filters, centrifuges, and stabilizers allow for the production of shelf-stable beer without extended lagering.

136. What are typical pitch rates for today's lager beers?

In modern lager breweries, the pitch rate is 1–1.25×10^6 cells per milliliter per degree Plato. A 12°P beer is pitched at 12–15×10^6 and high-gravity beers at 20×10^6. The specialty brewer must be cautioned that the actual required pitch rate should not be based solely on a rule of thumb, but on yeast viability and vitality. Other considerations when choosing pitch rates are the fermentation temperatures, yeast strain, and desired beer flavor characteristics (see earlier discussions on yeast and pitching).

Excessive pitch rates lead to a rapid start to fermentation but can cause the beer to taste thin and empty. Yeast growth is limited by available wort nutrients and oxygen so that fewer new cells are produced, leading to a degradation of the culture during serial repitching. A lower pitch rate, however, leads to slower fermentation and possibly a full, estery beer flavor and, of course, more cell growth.

Actual fermentation parameters should be based on the results of practical experiences in a given brewery and not on what works in other breweries.

137. What are aging and lagering?

Following primary fermentation, the young beer is matured at cooler temperatures. This process can last from one day to eight weeks depending on the beer style.

"Lagering" refers to extended aging of beer, particularly bottom-fermented beer, at low temperatures and generally involving a second fermentation. Some top-fermented beers such as kölsch and alt beers also require a period of lagering and are sometimes referred to as "top-fermented lager beers," but most often when brewers speak of "lager beer" they are referring to bottom-fermented beers (in German, *lager* means "to store"). Ruh storage refers to a period of aging without an active second fermentation (*ruh* comes from a German word meaning "to rest").

The purpose of aging is to promote the following:

a. Flavor maturation, eliminating the undesirable flavors and creating the desirable ones

b. Clarification through sedimentation of yeast and other insoluble material

c. Chill stabilization by promoting the creation of protein-tannin complexes that can be removed by sedimentation or filtration

d. Development of carbonation, if a secondary fermentation is employed; some brewers flush the ruh beer with CO_2 gas to help scrub out undesirable volatiles, especially sulfur compounds

138. What methods are used today for lager fermentation and aging?

There are numerous variations on lager brewing techniques, but they can be divided into two categories: traditional and modern methods.

139. How do traditional methods and modern methods differ?

Although modern fermentation takes place at higher temperatures, both systems start out in similar ways. However, at peak fermentation ("high kräusen"), the processes diverge. In the traditional processes, the brewer begins to slowly cool the beer in preparation for a long secondary fermentation. With modern processes, the brewer lets the fermentation temperature ramp up in order to facilitate rapid reduction of diacetyl.

140. What is an example of a traditional method of lager fermentation?

Primary fermentation.   Wort is cooled to 45°F (7°C), aerated to a DO content of 8 ppm, and pitched (*Figure 1.32*). As the beer ferments, the temperature rises, and after 1–2 days is attemperated to 48°F (9°C). After 4–5 days, the yeast growth phase is complete; the yeast cell count peaks; and the apparent gravity is about half of the original gravity. The beer is slowly cooled, lowering its temperature by a maximum of 4°F (2°C) per day, until it reaches 39–41°F (4–5°C). The beer should contain 0.8–1.2°P fermentable extract and 5 million to 15 million cells per milliliter. A modification of this process starts with a cold-wort temperature of 52°F (11°C), rising to 56°F (13.5°C) (*Figure 1.33*).

Secondary fermentation.   If a unitank process is used, the tank is bunged to develop natural carbonation. Yeast for repitching should be

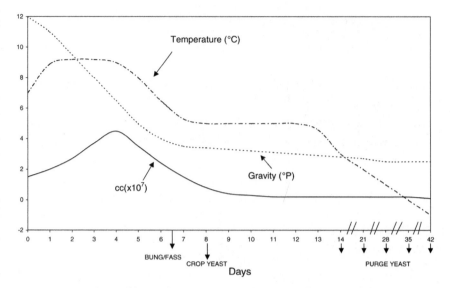

Figure 1.32. Traditional lager fermentation and aging cycle. cc = cell count (number per milliliter).

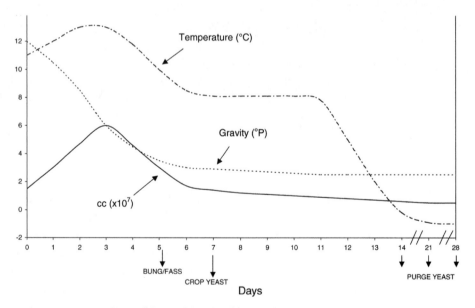

Figure 1.33. Traditional lager fermentation and aging cycle with elevated starting temperature. cc = cell count (number per milliliter).

removed no more than three days after bunging. Further yeast should be bled from the cone at least once per week during lagering and discarded.

Table 1.6. Fermentation and lagering parameters for traditional lager beer

	Minimum	Maximum
Primary fermentation	5 days	12 days
Lagering	3 weeks	6 weeks
Pitch temperature	42°F (6°C)	52°F (11°C)
Peak temperature	46°F (8°C)	56°F (13°C)
Peak cell count (original gravity 12°P)	30×10^6	65×10^6
Temperature at bung	39°F (4°C)	48°F (9°C)
Bunging pressure	6 psi	15 psi

Figure 1.34. Typical Czech lager fermentation cellar. Open fermenters are slowly being phased out in modern breweries but are still used in some traditional plants.

If a two-tank process is used, the beer is fassed to a lager tank when the apparent gravity is about 1°P above final attenuation. The lager tank is left venting for 1–2 days and then is bunged.

The beer can be held at 39–41°F (4–5°C) until the VDK level is below 0.15 ppm and is then slowly cooled to 30°F (−1°C). Total lager time is 1 week for every 2–3°P original gravity; 4–6 weeks is normally sufficient for a 12°P beer. Active fermentation takes place during the first half of lagering as the fermentable extract is slowly depleted. During the second half of lagering, there is little fermentation but the beer flavor improves due to continued yeast activity. Excessive lagering, especially at elevated

Figure 1.35. Traditional lager cellar in a large Czech brewery, consisting of single-shell, long, narrow lager tanks stacked in an air-cooled lager cellar. Note the brass bunging apparatus on the front of each tank.

temperatures, leads to yeast autolysis and loss of beer quality. *Table 1.6* shows parameters for traditional lager beer fermentation and aging.

Figures 1.34 and *1.35* show traditional open fermenters and a lager cellar in a Czech brewery.

141. How is the fermentation temperature influenced by fermenter design?

If primary fermentation temperatures below 48°F (9°C) are used, the two-tank cellar system is preferred. Use of traditional cold-fermentation temperatures of 42–46°F (5.5–8°C) with the unitank system often produce inferior beers. However, excellent lager beers can be produced in unitanks when mid-range temperatures of 48–56°F (9–13°C) are employed.

142. What is the significance of the rate of cooling after peak cell count in traditional fermentation?

The cooling rate at the end of the yeast growth phase is particularly important if the beer is to be transferred to a separate lagering tank. The purpose of a controlled temperature drop during the second half of fermentation is to promote yeast flocculation and to slow the fermentation rate so there will be enough fermentable extract for a proper secondary fermentation and natural carbonation. If the beer is cooled too fast, the yeast might be "thermally shocked," or even be forced out of solution, leaving insufficient healthy yeast for lagering and natural carbonation. If the beer is too warm at fass, too much yeast could be carried over. The ensuing rapid secondary fermentation is harmful to beer flavor and can lead to autolysis.

143. What happens to the yeast during long cold-lagering?

Yeast is important for flavor development in extended lagering. Traditional cold, long lagering produces a full-flavored beer with improved foam stability. The reasons for this are poorly understood but may be partly due to changes in the yeast as the beer ages. During cold-lagering the yeast moves through the following stages. During the first half of lagering, yeast continues a slow secondary fermentation without growth. During the last half of lagering, fermentation is complete but maturation continues and yeast cell membrane degradation releases metabolic products such as keto acids, amino acids, nucleotides, inorganic phosphates, and glycerol. These compounds increase the palate fullness of beer and can enhance flavor and foam. At this point, lagering should be terminated. If lagering continues, especially at elevated temperatures, the yeast die and release cellular enzymes, lipids, and other compounds deleterious to beer flavor and foam. A rise in beer pH indicates that autolysis is occurring.

144. What is an example of a modern fermentation method?

Primary fermentation. The wort is cooled to 48°F (9°C), aerated until its DO content is 8 ppm, and pitched (*Figure 1.36*). After 1–2 days, the fermenter is attempered to 52°F (11°C). Peak cell count is reached

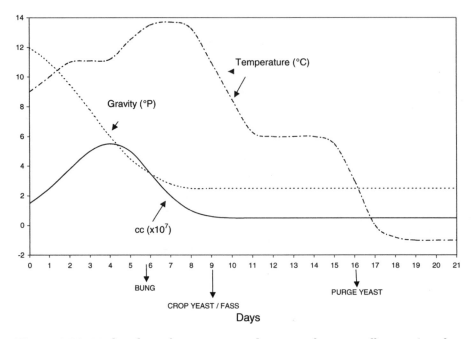

Figure 1.36. Modern lager fermentation and aging cycle. cc = cell count (number per milliliter).

after 3–4 days, and yeast growth is complete; the apparent gravity is roughly half the original gravity. The beer is then allowed to "free rise" to 56°F (13.3°C) for a "diacetyl rest," where it is held for 1–3 days until the VDK level is below 0.15 ppm. The difference between the temperature at pitching and at diacetyl rest is best kept below 9 degrees F.

Conditioning. If a unitank is used, the tank should be bunged against 14 psi (less in deep unitanks), with about 1°P fermentable extract remaining. Yeast for repitching should be harvested no more than three days after terminal gravity is achieved.

If a two-tank process is used, the end-fermented beer is transferred to a conditioning tank, so that yeast can be harvested from the fermenter. Because there is little fermentable extract left, the brewer must be careful to minimize air pickup during the transfer and may benefit from purging the transfer piping and conditioning tank with inert gas.

The beer is slowly cooled to 43°F (6°C) and held for 4–6 days. It is then cooled to 30°F (–1°C) for at least two days before filtration and carbonation. The entire process of fermentation and conditioning typically takes 21 days. If this method of fermentation is used without subsequent

Figure 1.37. Fermentation cellar in a medium-sized German brewery. (Courtesy of Christian Gresser GmbH)

Table 1.7. Fermentation and storage parameters for modern lager

	Minimum	**Maximum**
Primary fermentation	4 days	7 days
Lagering	1 day	28 days
Pitch temperature	46°F (8°C)	55°F (13°C)
Attemperation temperature	50°F (10°C)	64°F (18°C)
Peak cell count (original gravity 12°P)	30×10^6	80×10^6
Diacetyl rest temperature	52°F (11°C)	68°F (20°C)

kräusening, it is, of course, not possible to fully carbonate the beer naturally by bunging the tank.

Warmer fermentation procedures are common—for example, pitching at 2×10^6 cells per milliliter per degree Plato and starting fermentation at 54°F (12°C) and rising to 63°F (17°C), with a 2-day diacetyl rest. In this case, the total process time can be as short as eight days!

Table 1.7 shows parameters for modern lager beer fermentation and aging. *Figures 1.37* and *1.38* show cellars in modern German breweries.

Figure 1.38. Typical modern brewery tank farm and pipe fence. Such vessels can be used for either fermentation or aging. Although not common in Europe, most North American specialty brewers use such vessels as unitanks. (Courtesy of Christian Gresser GmbH)

145. Is there an advantage to cold fermentation temperatures?

Traditional brewers prefer to ferment below 55°F (13°C). Lower temperatures are believed to produce fuller, mellow beers with superior foam. Cold fermentations limit the production of many fermentation by-products, with the exception of some esters that may actually increase. In the Czech lager brewing tradition, conducting primary fermentation below a maximum of 48°F (9°C) is preferred, in part to keep peak fermentation VDK levels at or below about 0.3 ppm. Thus, after lagering, the final beer VDK level will be below 0.15 ppm. The modern method of warmer fermentation temperatures produces beer with a lighter flavor and lower VDK, which results in the good drinkability required by today's beer drinkers. The specialty brewer must, however, choose fermentation procedures based solely on empirical evidence gathered in his or her own brewery. As brewing processes and raw materials change, traditional procedures may not be applicable.

146. Why is fermentation started at a low temperature and allowed to rise to a higher temperature?

Wort is rich in nutrients, so to discourage microbiological contamination, brewers start fermentation at cooler temperatures until yeast has a chance to "take hold" and before jacket cooling begins. Empirical evidence indicates that a cooler start produces a smoother beer.

Yeast growth occurs during the cooler first half of fermentation; thus, unwanted fermentation by-products are better kept in check.

If fermentation is started too warm, the process can move too quickly, exhausting the extract needed for secondary fermentation and leaving too many yeast cells in suspension and not enough yeast to recover for subsequent fermentations.

147. What is Drauflassen?

Drauflassen is a German brewing term that roughly translates as "to give an addition, top up, or double." This "doubling" refers to the addition of fresh wort to fermenting beer. By topping actively fermenting beer in the growth phase with fresh wort, the amount of healthy yeast cells produced will be increased. If a brewer has only enough yeast to pitch one brew but wishes to make more beer quickly, it is better to use the Drauflassen process then to underpitch. Brewers sometimes face a shortage of yeast during yeast propagation, e.g., if one is starting up a cellar after a prolonged shutdown or if a yeast strain with poor flocculation characteristics is used. An example of the Drauflassen process is as follows. A volume of wort is pitched at the proper rate and aerated. It is allowed to ferment for 24–48 hr, depending on the multiplication rate of the yeast, to achieve, for example, more than 20×10^6 cells per milliliter. It is then topped with a similar volume of aerated wort, with no added yeast, so that when the tank is full, there is sufficient viable yeast for proper fermentation. This fermenter can then be "split and topped" at high kräusen as needed to quickly start further fermentations.

148. Why are dark lagers traditionally fermented at warmer temperatures than pale beers?

Traditionally, dark lagers are fermented at a temperature a few degrees warmer than pale beers because dark beers are more difficult for yeast to ferment. During the kilning of Munich-type malts at around 212–220°F (100–105°C), a certain amount of amino acids as well as mono- and disaccharides are utilized for browning reactions, leaving less "starter sugars" in the wort. Use of large amounts of dark-roasted malts has been shown to decrease attenuation and fermentability. This is a result of lower levels of fermentable sugars and amino acids in dark wort samples as well as the inhibitory effects of the wort Maillard compounds on yeast metabolism (Coghe et al., 2005).

149. What is kräusening?

In brewing, *kräusen* (a German word meaning "ruffles") refers to the curly white froth formed on top of actively fermenting beer. Kräusen beer is young, fermenting beer at 10–30% attenuation. In the case of lager beer, it is 1- to 3-day-old fermenting beer with the characteristic foam cover, or kräusen, of primary fermentation. Kräusen beer may have an apparent gravity of 6.5–9°P and contain $25–50 \times 10^6$ cells per milliliter.

The kräusening process is a method of secondary fermentation to ensure proper flavor maturation and natural carbonation. In the kräusening process, beer is fermented to completion and cooled to about 42°F (5.5°C) before being pumped to the lager tank. This beer has no residual fermentable extract available for secondary fermentation. Therefore, it is topped up with 10–15% by volume of kräusen beer from the fermenting cellar, which increases the apparent gravity by about 1°P and causes renewed fermentation. The lager tank is left venting for 1–3 days and then bunged. Although the actively fermenting yeast causes renewed development of metabolic by-products, such as diacetyl and acetaldehyde, the healthy yeast ultimately lowers the final levels into the acceptable range, so that beers made by the kräusen process can actually mature faster than traditional secondary-fermented beers. The kräusen process takes 3–6 weeks, as the beer slowly cools from 42°F (5.5°C) to 30°F (–1°C). Kräusening at cold temperatures sometimes leaves a small amount of residual fermentable extract (0.3–0.4°P) in the finished beer, which might be beneficial in some beer styles.

To promote vigorous reduction of immature beer flavors, lager temperatures as high as 52°F (11°C) are sometimes used.

Kräusening is particularly beneficial when the house yeast strain is very flocculent. Strongly flocculent yeast makes traditional lagering difficult because most of the yeast drops out of solution in primary fermentation, leaving elevated diacetyl levels and insufficient yeast for secondary fermentation. Kräusening also helps the beer resist "overaging" during cold storage because the active kräusen yeast absorbs FAN and fatty acids left in the beer after the main fermentation. Kräusen yeast generally excretes little fatty acids and proteases, thus promoting a more stable final beer foam (Dr. Ludwig Narziss, *personal communication,* August 2004).

In breweries adhering to the *Rheinheitsgebot,* kräusening—or using CO_2 collected from fermentation—might be the only option available for carbonation of problem batches if a lager tank should not fully carbonate due to insufficient secondary fermentation or loss of pressure due to a leak.

150. Is extended cold-lagering necessary with modern fermentation methods?

With modern fermentation techniques, there is little extract and yeast left in the beer during aging, so long lagering is not needed and may actually harm the beer. Long lagering requires the presence of fermentable extract, which is nearly completely depleted during the diacetyl rest, so a high yeast concentration remaining in the beer would autolyze to the detriment of the beer flavor and foam. Extended lagering is expensive, because inventory is tied up for an extended period. With the use of in-line beer coolers, stabilization aids, centrifuges or filters, and pinpoint carbonators, aging can be cut to less than 2 weeks. In the summer peak sales months, some brewers might "lager" for only 1 day. The lager tank, in this case, really serves only as a buffer tank to supply the finishing cellar! Some successful international lagers are produced in this manner, and the beer is of high quality, although possibly lacking in complexity.

Long cold-lagering is a cornerstone of traditional brewing and is required for proper flavor maturation as long as sufficient extract and healthy yeast remain in the beer for secondary fermentation. Traditional cold fermentation and extended secondary fermentation result in beers with full, round, complex flavors, as well as excellent foam stability. Nonetheless, beers made in this fashion often contain higher levels of VDK. Although some brewers boast of "90-day lagering," most beers require a maximum of 4–8 weeks lager time, or even less in low-hopped beers. Longer lagering is usually reserved for strong beers.

Although brewers have always been influenced by others, blind adherence to tradition without a clear understanding of its purpose does not lead to success. Successful brewers rely rather on the use of scientific methods to ensure the best manner of production in their particular brewery, regardless of what works for others.

151. What are the critical quality control points of lager fermentation?

Pitch rate/yeast viability
Starting temperature
pH drop during the first 24 hours of fermentation
pH of beer after maturation
Attenuation limit
Peak cell count
Gravity and temperature at peak cell count
Cell count at bung or fass
Gravity and temperature at bung or fass
VDK level before cooling to lager temperature
Final attenuation
Growth ratio (yeast recovered; yeast pitched)
Brink yeast temperature during storage
Decline of extract during aging (important when using traditional
    lagering methods)
Beer taste

152. What is *brandhefe*?

Brandhefe (sometimes mistakenly referred to as *braunhefe, brannt-hefe,* or *brauhefe*) is a brown, gummy material consisting of dead yeast, cold break, and oxidized foam (in German, *brandhefe* means "burnt yeast"). It forms on top of the fermenter foam cover and is left behind on the tank walls and ceiling as the foam cover collapses. *Brandhefe* has a harsh bitter taste and, if not removed from milder beers, interferes with beer drinkability. *Figures 1.39–1.42* show the various stages of foam formation during a lager fermentation in an open fermenter. *Figure 1.43* shows the *brandhefe* ring on an emptied fermenter.

153. How can brewers remove *brandhefe* during fermentation?

Brandhefe can easily be skimmed off an open fermenter. China clay, an inert white paste, can be painted on the upper walls of an open fermenter before filling (*Figure 1.44*). It forms a chalk layer that makes re-

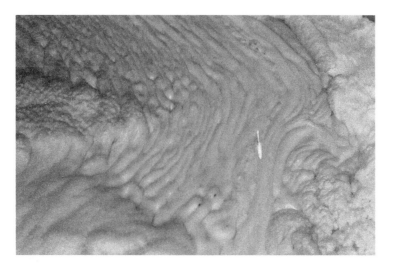

Figure 1.39. Initial fermentation heads.

Figure 1.40. Rocky "high kräusen" fermentation heads.

moving the sticky *brandhefe* easier, as the chalk with adhering *brandhefe* easily washes off the tank wall.

It is more difficult to remove *brandhefe* from beer in a closed fermenter. The simplest solution is to fill the tank to such a level that the foam cover rises and contacts the top of the tank. The *brandhefe* will stick to the tank walls and ceiling—a slight fermenter foam-over can also be beneficial to a smoother final beer flavor. Of course, such methods require sanitary vessel design and proper CIP methods.

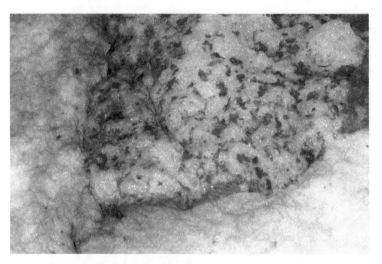

Figure 1.41. *Brandhefe* formation on fermentation heads.

Figure 1.42. Final heads on finished fermentation.

154. What is hefeweizen beer?

Hefeweizen is beer brewed with *weizen* ("wheat") and containing *hefe* ("yeast"). It is one of the most popular styles of beer in Germany. Hefeweizen beers are sometimes also called Hefeweiss or simply Weiss Bier (*weiss* means "white" in German—the name originated in the Middle Ages, when wheat beers were noticeably paler than the typical brown beers brewed with 100% barley malt).

Figure 1.43. Fermentation residue on an emptied open fermenter.

Figure 1.44. China clay on an open fermenter to assist in later cleaning.

This type of weiss beer should not be confused with the less popular, low-gravity Berliner Weiss, a different style of pale wheat beer brewed in Berlin and noted for its refreshing but intense lactic acidity.

Hefeweizen beer is made from a blend of malted wheat (making up at least 50% of the grist) and malted barley using a unique strain of top-fermenting yeast. Generally, hefeweizen beers are not highly hopped (10–

14 international bitterness units). There are numerous subcategories of hefeweizen, with colors ranging from pale to dark and strength ranging from low-calorie to double bock. Hefeweizen beer is traditionally bottle fermented and highly carbonated—for example, to more than 3.10 volumes of CO_2 (0.60%) for draft versions and to more than 3.60 volumes (0.70%) for bottles. Filtered wheat ales, called Kristall Weizen, sometimes contain more than 3.9 volumes of (0.75%) CO_2. Most standard North American beer bottles are designed for a maximum 3.0 volumes of CO_2, so special bottles must be used at such high pressures. Most North American draft systems are not suitable for such high CO_2 levels, and therefore the brewer is cautioned to avoid CO_2 levels above about 2.80 volumes in draft beer.

Some modern examples of hefeweizen beer are actually end-fermented in the brewery and do not undergo a true bottle fermentation. Because of the required high final carbonation, such breweries need pressure vessels suitable for at least 30 psi.

155. What are the unique characteristics of hefeweizen beer?

Traditional German hefeweizen should have the following characteristics:

a. Soft, sweet, bready flavor.   This flavor is the result of using malted wheat and a low hopping rate, together with the presence of yeast in the bottle.

b. Spicy aroma.   An aroma reminiscent of cloves and sometimes wood smoke or vanilla is due to phenolic fermentation by-products. Phenol chemistry is complex and not completely understood, but the most important phenol in weizen beer is believed to be 4-vinyl guaiacol. The precursor of this compound is ferulic acid, which originates mostly from barley malt and, to a lesser extent, from wheat malt. Mashing at 95–130°F (35–55°C) fosters the hydrolysis of ferulic acid. The unique strains of yeast used to make weizen beers are adept at utilizing ferulic acid to produce 4-vinyl guaiacol and its spicy aroma. Although mashing at lower temperatures promotes ferulic acid production, the increased proteolysis results in less ester development during fermentation.

c. Fruity aroma.   An aroma reminiscent of bananas is due to esters such as isoamyl acetate, created by yeast during main fermentation and bottle conditioning.

d. High carbonation and excellent foam stability. The signature pour into tall glassware produces a thick head of foam.

e. Beer haze. Most haze particles will eventually drop out of solution due to gravity. It is therefore not possible to create haze that is stable for much more than two months unless special measures are taken. A stable haze has become a critical quality parameter for modern unfiltered wheat beers.

156. What methods can be used to create a stable haze in beer?

Sources of haze can be yeast, protein/polyphenol complexes, starch, or other additives.

a. Yeast. In traditional hefeweizen beer, the presence of viable yeast in the bottle is necessary for development of carbonation and fine flavor. But as yeast settles, it will "drag" haze-forming proteins and carbohydrates out of solution. If a stable haze is desired, it is imperative that a powdery strain be used for bottle refermentation, and one must accept the fact that, after some time, the yeast will drop out of solution. Therefore, some modern hefeweizen beers are completely carbonated in tanks and centrifuged before bottling so that they contain almost no yeast.

b. Protein/polyphenol complexes. The optimum particle size for stable haze appears to be between 0.2 and 1 micron in diameter (Brandl and Englmann, 2001). Although small particles will produce less haze than larger particles, they drop out of solution at a much slower rate. It is not only the size of the protein molecules but also the type of proteins that impact haze. It is important to break down, alter, or remove large proteins before bottling so that they do not bind and precipitate with polyphenols. Proteins containing high concentrations of the amino acids proline and glutamic acid seem to promote a more stable haze (R. Mussche, *personal communication,* May 2005).

A few suggestions can be offered to optimize protein levels.

1. Types of grains. Barley and wheat varietal choices, growing conditions, and malting methods have a great impact on subsequent beer haze stability. Malted wheat is preferred over unmalted wheat (Delvaux et al., 2004). Sometimes a haze problem can be resolved simply by changing malt suppliers.

2. Amount of wheat used. Some researchers believe that it is best to use as much malted wheat as possible (M. Herrmann and C. Schwarz, *personal communication,* May 2005), while others counter that

2–30% malted wheat is preferred (Delvaux et al., 2003; R. Mussche, *personal communication,* May 2005). The flavor impact due to the amount of wheat used must be considered.

3. Malt protein content.   It is believed that high-protein malts are preferable and that increased protein breakdown during malting is beneficial.

4. Kettle boil.   A shorter kettle boil resulting in a total coagulable nitrogen level greater than 3 mg/100 ml is helpful (Brandl and Englmann, 2001).

5. Use of centrifuge.   Removing large protein/polyphenol complexes and yeast while leaving smaller molecules in suspension can create a stable haze. Also, the shear forces generated during centrifugation may help produce stable haze particles.

6. Flash pasteurization.   The rapid heating process of flash pasteurization alters proteins so that they stay in suspension longer. Little is understood about this process. Typical holding temperatures are 165–169°F (74–76°C) for 20 s. Sometimes the beer is pasteurized right after primary fermentation, which may be preferable to pasteurizing at a later stage because, during aging, much of the protein haze would be lost due to sedimentation (Brandl and Englmann, 2001). As long as DO levels are minimal (0.02 ppm or less), there is no detrimental effect from flash pasteurization even in the presence of suspended yeast.

c. Starch.   Starch remaining in beer will produce haze. It is possible to manipulate the mashing process so that not all of the malt starch is broken down, thus producing a wort with a "slight iodine-positive reaction." Wheat flour can also be added to the brew kettle.

d. Additives.   Natural polysaccharide gums or small molecular weight gallotannins are sometimes added to wheat-based beers and are extremely effective at generating a stable haze.

157. What is unique about weizen beer yeast strains?

Weizen beer yeast is a special type of top-fermenting yeast, belonging to the taxonomic species *Saccharomyces cerevisiae.* (Some brewers mistakenly refer to this yeast as *Saccharomyces delbrueckii,* but such a yeast does not exist! The confusion may lie in the fact that *Lactobacillus delbrueckii,* a bacterium, is important in the fermentation of other types of wheat beer, including Berliner Weiss and some Belgian wheat beers. Low-hopped hefeweizen beer is susceptible to bacterial spoilage, so *L. delbrueckii,* like any bacteria, is an unwanted intruder in a weizen beer cellar.)

Weizen beer yeast produces high levels of spicy phenols and certain esters. Although brewers have manipulated these yeasts for use in modern vertical tanks, best results are always achieved in shallow, open fermenters. Because of the specialized characteristics of this group of yeasts, there are actually few strains available. Among these strains, slight variations are seen in the ability to produce the desired spicy ester character as well as flocculation. Some brewers have reported large differences in genetic stability among strains, especially with repitching from cone-bottomed fermenters. When mishandled, weizen yeast produces excessive levels of H_2S and even diacetyl.

Weizen beer strains originated in Germany and can be purchased from yeast banks there. The most popular strain is Weihenstephan #68.

158. What are the special considerations for the fermentation of hefeweizen beer?

Successful fermentation of hefeweizen yields a beer with a balanced flavor—soft, slightly sweet, spicy, estery, and bready. Excessive bitterness should be avoided by keeping the hop rate low and ensuring that excess yeast is removed from the beer once fermentation is complete. This can be extremely difficult when using nonflocculent strains (such as Weihenstephan #127) in unitanks. Most successful German weiss beer brewers use either traditional open fermenters for top cropping (*Figure 1.45*) or a centrifuge when using modern vertical tanks. Flocculation aids such as isinglass can also be used. Using powdery yeast without a centrifuge or filter can lead to excessive yeast carryover to the package. With open fermenters, yeast can be harvested and repitched for hundreds of generations. But if closed vertical tanks are used, it is difficult to maintain genetic stability and even to harvest sufficient yeast for repitching after five or 10 generations, especially when vessels are over 12 ft deep. In such instances, some brewers start with a fresh culture for every batch or simply add about 10–25% actively fermenting kräusen beer to the aerated wort to induce fermentation. In such cases, because the yeast does not flocculate well, the yeast is never truly harvested.

The development of the spicy, phenolic character is a result of adequate development of the necessary precursors during mashing and the selection of the appropriate yeast strain. The unique ester profile of this beer is strain-dependent but can be influenced by fermentation parameters. (See the discussion on esters.) It has been suggested by some brewers that ester levels can be increased by limiting wort aeration. Although

Figure 1.45. Weiss bier fermenter, showing the self-skimming chute. (Courtesy of Christian Gresser GmbH)

this is sometimes true, depending on the yeast strain, the practical brewer must be careful not to starve the yeast of oxygen, which can result in a faulty fermentation, significant off-flavors (sulfur compounds), as well as bacterial attack. It is, therefore, critical to inject sufficient oxygen for yeast health while avoiding excess aeration. (See the earlier discussion of wort aeration.)

159. What are example of typical hefeweiss fermentation and aging programs?

a. Fermentation and lagering parameters for hefeweiss beer

Primary fermentation time	3–5 days
Pitch rate	$0.3–1 \times 10^6$ cells per milliliter per °P
Fermentation temperature	54–77°F (12–25°C)
Lagering time	0–35 days
Lager temperature	39–55°F (4–13°C)

b. Primary fermentation. A typical example might include pitching at 7×10^6 cells per milliliter, with fermentation starting at 65°F (18.5°C) and rising to a maximum of 70°F (21°C). Primary fermentation is complete after four days. The flocculated yeast is cropped in the normal fashion.

c. Secondary treatment. There are many ways to handle hefeweiss beer once primary fermentation is complete. Four examples are outlined below.

1. Traditional No. 1. Like English cask ale brewing, hefeweiss brewing can be simple. In some small breweries, the beer is transferred directly from the fermenter to a mixing tank. Sometimes a coarse filtration is used to remove trub. A dose of 10% by volume of *speise* (wort) is added. A dose of fresh yeast is not needed, as carryover is sufficient. After mixing, the beer is bottled. Small breweries using this process make very special beers but, due to the high yeast carryover and the lack of tank aging, these beers can be variable in flavor and stability.

2. Traditional No. 2. The beer is transferred to an aging tank, cooled to 39–54°F (4–12°C), and held for seven days or more so that the flavor matures and the yeast settles. The beer is then sometimes coarse filtered or centrifuged. Fresh weiss yeast and speise are added to the flat beer, and it is bottled. Because of the secondary tank aging, controlled yeast carryover, and bottle fermentation, this process produces clean and round beer flavor.

3. Modern No. 1. After primary fermentation, the beer is centrifuged and sometimes flash pasteurized without secondary aging. Fresh yeast as well as speise is added, and the beer is bottled. This process also produces fine beer flavor as long as bottle aging is handled properly.

4. Modern No. 2. The beer is transferred to an aging cellar with or without centrifugation or flash pasteurization. It is cooled to 44–50°F (7–10°C). Lager yeast kräusen is added, and the beer is held for 7–10 days to develop carbonation. It is then cooled to 43°F (6°C) or for less for 3 weeks. The lager yeast will settle, leaving a stable nonyeast haze. Lager yeast ferments at cold temperatures; thus, sufficient carbonation is developed in a 30-psi tank. If weiss yeast is used, a higher temperature is required, so the aging tank must withstand 45 psi. The finished beer is bottled without refermentation. Bottle refermentation can sometimes be problematic; thus, the brewer has greater control through tank fermentation. This method produces a beer with long-term haze stability and consistency but, unfortunately, lacking the unique characteristics and benefits of bottle fermentation.

d. Bottle refermentation.   At least 100,000 cells per milliliter of yeast is required, although some weiss beers contain more than 10,000,000 cells per milliliter. After the beer is bottled or kegged, it is held in the brewery to referment for 3–14 days at 68–77°F (20–25°C). The bottles or kegs can be held for a further 7–14 days at around 50°F (10°C) to finished. Strong beers may need as much as 42 days to fully mature.

160. Because weiss beer is bottle-fermented, is it necessary to minimize oxygen pickup during packaging?

In most cases, the brewer is advised to follow standard practices of "air discipline" during postfermentation handling and packaging. There is very little yeast growth during bottle refermentation, so little oxygen is required, especially if the yeast slurry has been properly prepared. Excess oxygen will simply contribute to long-term staling.

However, if the base beer is of high original gravity (>18°P) or is of dark color, or if the yeast strain is not particularly vigorous, a maximum of 2 ppm DO in the bottled beer may be of some benefit (Derdelinckx et al., 1992).

Ale Fermentation and Conditioning

161. How were ales produced before refrigeration?

After boiling, the wort was cooled utilizing fans, natural convection, or cold water in an open cooler or "refrigeratory worm" to 50–65°F (10-18°C). Yeast slurry was added—about 10 gallons per 30 bbl (also referred to as "store")—to the gyle-tun and fermentation proceeded for about 8–12 days. The fermentation temperature would gradually rise to 72°F (22°C) in small vessels utilizing ambient cooling. After fermentation slowed, the beer would be racked into small barrels, to be cleansed and conditioned, the yeast being allowed to flow out the open bunghole. In warm weather, racking into small barrels took place sooner to help control fermentation temperature. The casks were kept topped up with beer to ensure that the yeast was allowed to separate and to minimize air contact. The beer was fined with isinglass and delivered to the publican to allow for final conditioning and clarification. This type of ale production was typically limited to cool climates or seasonal production, although floating copper "swimmers" filled with ice, when available, could be used in the fermenters to control fermentation temperatures.

162. How have ale brewing practices changed in the last 150 years?

Technological advances in brewing have allowed for much more consistent and efficient production of ale, most notably from the following process improvements:

a. A better understanding of brewing chemistry has allowed the production of more-uniform and better-quality wort.

b. Artificial refrigeration and automated temperature control have allowed much more consistent fermentation and flavor profiles.

c. The use of modern fabrication materials and methods has allowed larger batch sizes and operational efficiencies.

d. Improved yeast handling and pure culture isolation have provided improved consistency.

e. A better understanding of the positive and negative effects that oxygen plays in pre- and postfermentation beer quality has improved flavor stability.

f. Innovations in beer treatment have allowed for a much more flavorful and physically stable product.

163. What is the classic ale fermentation process?

Traditional cask ale brewers brew for the on-premise pub trade. This trade demands lower-alcohol "session beers"—typically 3.8–4.2% alcohol by volume, with pH in the range of 3.7–4.1. The shelf life for cask beer is typically 4–5 weeks (Atkinson et al., 1985, p. 43). The cellaring process is warm and quick; it is not unusual for a beer to be brewed and stillaged in the pub ready to serve within eight days.

Yeast is typically added directly into the tank or, in more modern breweries, injected in-line. Tanks are typically designed to receive a single brew and are filled with cooled wort from an open cooler, with some brewers utilizing a fishtail spray when pumping into the top of the fermenter to elevate dissolved oxygen levels in the wort. *Figure 1.46* shows an open fermenter filling from the bottom. It is more likely today that a closed heat exchanger with in-line aeration will be used in place of the open cooler, due to the difficulty in cleaning and sterilizing this equipment during this critical step of the process.

Pitching temperature is typically 59–62°F (15–17°C), climbing to 67–70°F (20–22°C). Cold-water cooling coils (*Figure 1.47*) or jackets may be fitted to control fermentation temperature either manually or with

Figure 1.46. Filling the open fermenter from the bottom "down tube." (Courtesy of St Austell Brewery)

Figure 1.47. Cooling coils in an open ale fermenter.

thermostatic controls, although small, shallow tanks may rely on ambient room cooling. Collected heads should be clean and bright with no trub carryover. The dirty yeast heads are typically skimmed early in the fermentation, allowing cleaner yeast to form for later harvest.

Figure 1.48. Adjustable collection funnel for skimming yeast heads.

The top-cropping yeast can be manually scooped off the surface, or breweries may utilize tanks equipped with permanently mounted yeast-collection "inverted funnels" that can be raised or lowered into the fermenter to skim and harvest the yeast (*Figure 1.48*). Other designs utilize tanks that are equipped with a side chute at a predetermined or adjustable height that allows the head to self-skim. Other breweries may be equipped with overhead vacuum connections that allow the yeast to be sucked off to a central yeast-collection system. A modified shop vacuum was observed in use for this purpose at one small brewery.

Some brewers utilize the practice of pumping over, or rousing, during the fermentation to help keep very-flocculent yeast strains in suspension and dissolve additional oxygen to make up for possible insufficient initial saturation. After fermentation is complete (typically in 3–4 days, but high-gravity beers may require additional time) and skimming or vacuuming has removed the yeast, the beer is drained or pumped from the vessel, leaving behind additional sedimented protein and dead yeast. The beer is rarely filtered but can be lightly primed with sugar or transferred with a small amount of residual extract, fined with isinglass, and then racked directly into casks for final conditioning at 54–57°F (12–14°C) and dispensing with little additional aging. *Table 1.8* shows cellar parameters for traditional ale fermentations and *Figure 1.49* shows the

Table 1.8. Fermentation parameters for traditional ale

	Minimum	Maximum
Primary fermentation	3 days	6 days
Secondary fermentation	0 days	7 days
Pitch temperature	58°F (15°C)	72°F (22°C)
Peak temperature	65°F (18°C)	77°F (25°C)
Pitch rate	5×10^6	10×10^6
Peak cell count	40×10^6	65×10^6

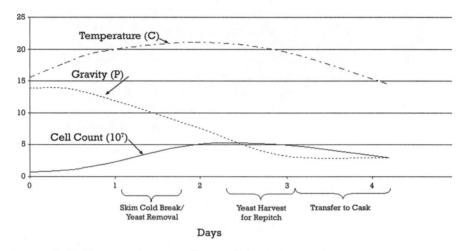

Figure 1.49. Fermentation curve for a traditional ale over time.

fermentation curve over time. This type of product and other traditional top-fermented styles still exist today in certain parts of the world, notably the United Kingdom, but due to the difficulty in handling them and their typically short shelf life (five weeks), they have a limited market.

164. What is the classic ale fermenter?

The brewers usually utilize single, shallow (depth of 3–10 ft), open-top, rectangular or round, flat-bottomed tanks. Newer installations typically use stainless steel. Older installations may still have slate, copper, or wooden vessels, but many of these have been upgraded with polypropylene or other plastic liners. More-modern installations may have tanks with dome tops to assist with CIP cleaning and CO_2 collection and removal (*Figure 1.50*). Occasionally, movable CIP hoods are employed on older installations with tanks that were not designed with domes for CIP. *Figure 1.51* shows a classic open-fermenting cellar in a cask ale brewery.

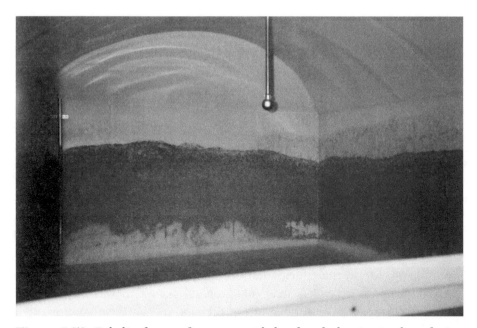

Figure 1.50. Poly-lined open fermenter with hood and cleaning-in-place device. Note the high degree of fermentation residue on the fermenter wall. (Courtesy of St Austell Brewery)

Figure 1.51. Open round fermenters in a British cask ale brewery. Note the side opening for yeast skimming.

165. What is the modern ale fermentation process?

Many variations and refinements have been made to the basic ale fermentation described above. Today, the demands of the marketplace put the same stringent requirements on the ale brewer for extended microbiological and physical stability and shelf life, as in modern lager production. Several distinct systems are now in use for ale production; systems utilizing smaller-sized, flat-bottomed fermenters with closed aging tanks, conical vertical fermenters, and unitanks are common.

Ales produced in shallow fermenters usually undergo a similar fermentation start as in cask ale production, although they are typically transferred to a secondary storage tank to complete fermentation and maturation. The temperature may be reduced to around 60°F (16°C) toward the end of fermentation (typically after 3–4 days) to help promote yeast flocculation. Modern ale fermentations typically involve vertical cylindroconical fermenters, which may take several brews to fill. *Figure 1.52* shows the process over time in a typical modern ale fermentation. It is desirable to transfer the beer with some remaining fermentation activity to help ensure that any oxygen pickup is absorbed by the active yeast. Some installations may use a plate heat exchanger to drop the tempera-

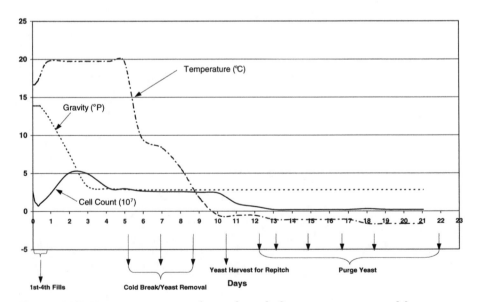

Figure 1.52. Fermentation curve for modern ale fermentation in conical fermenters.

Table 1.9. Fermentation parameters for modern ale

	Minimum	**Maximum**
Primary fermentation	3 days	5 days
Secondary fermentation	3 days	7 days
Cold stabilization	2 days	14 days
Pitch temperature	58°F (15°C)	72°F (22°C)
Peak temperature	65°F (18°C)	77°F (25°C)
Pitch rate	5×10^6	15×10^6
Peak cell count	40×10^6	65×10^6

ture during the transfer to the cellar. A warm conditioning or diacetyl rest may need to be incorporated at the end of fermentation, although some popular ales have a perceptible diacetyl character that is presumably part of their desired flavor profile. *Table 1.9* shows cellar parameters for modern ale fermentations.

Breweries utilizing conical fermenters have been able to achieve very similar flavor profiles compared to traditional fermentation systems. Changes in yeast handling are required since collecting yeast off the surface is not practical. Many ale strains appear to adapt to the change in environment and sediment into the cone instead of forming a top crop. Fermentation temperature and possibly pitch rate may need to be adjusted since the greater convection currents can accelerate the rate of fermentation considerably. Brewers utilizing a unitank process complete the steps described above in the same tank as that used for fermentation.

Some breweries utilize pressure vessels, allowing the beer to be transferred with sufficient residual extract to allow for bunging and the development of natural carbonation. After the transfer of the beer to storage and the completion of fermentation and maturation, the tank temperature is dropped to help sediment the remaining yeast. Ale maturation is dependent on yeast strain and fermentation temperature profile and varies from 1 or 2 days to several weeks, depending on the beer style and the brewer's desired flavor profile. Depending on what type of beer is being brewed and the required shelf stability, it may undergo a period of chill stabilization at 30°F (−1°C), or in some cases, it is filtered or fined and packaged directly after fermentation and maturation. Carbonation can be accomplished either by bunging, kräusening, priming, or injecting CO_2 afterward.

Figure 1.53. Conical fermenter in a specialty brewery. (Courtesy of Sierra Nevada Brewing Company)

166. How are conical fermenters used in modern ale production?

There are many variations of fermenting systems in use worldwide that have been developed and innovated over the years. The following designs are now widely in use for both ale and lager production.

In more-modern systems, the shallow primary fermenter has been replaced with a tall, vertical cylindrical tank with a steep-cone bottom (*Figure 1.53*), typically with a 60- to 90-degree cone angle (each side 30–45 degrees from vertical) and a closed top. This design allows for easy yeast collection and CIP cleaning. Steeper cone angles may help to ensure sedimentation and removal of yeast after fermentation, but steepness increases the height of the vessel and may decrease convection and mixing during fermentation and give less cooling area for the sedimented yeast. Most tanks have cooling jackets, typically with either glycol or direct-expansion refrigerant. Some cellar designs rely on circulating the beer through external cooling loops utilizing plate heat exchangers, but due to the lack of flexibility, increased chance of contamination, and complexity, these designs are not very common.

Since these tanks are not used for carbonating, they are usually designed for use at atmospheric pressure or a slight positive pressure. The cost of construction per barrel decreases dramatically as the tank size increases. Large tank size also provides the benefits of reduced cleaning, beer loss, installation piping, and control, as well as more efficient use of real estate. This has generally led to larger multifill fermenters that are sometimes located totally outside the cellar, or with just the cone and other necessary fittings protruding into the cellar. This design has created other new challenges in managing fermentation rate and profile since, during fermentation, the tall cylindrical design promotes active circulation, accelerating the rate of yeast activity. Since the economies of scale are so great with larger tanks, a trend developed for vessels with even greater capacity. Construction and shipping limitations resulted in very tall designs; some tanks have a fill height above 40 ft, resulting in hydrostatic pressure on the yeast of around 20 psi. Due to the stress on some yeast strains and other production problems caused by the high hydrostatic pressure, many of these tall tanks created operational problems. Most new tanks are now designed with a height-to-diameter ratio of less than 2.5:1 and with more moderate capacities.

After primary fermentation ceases and the sedimented yeast is collected, the beer is then transferred to either another vertical tank or a horizontal tank to complete maturation and aging. Some installations incorporate a plate heat exchanger to cool the beer on the way to storage. Carbonation can be accomplished either by bunging, adding fresh wort, priming, or with CO_2 injected after aging and clarification. A separate starting tank may also be employed with this system.

A further evolution in cellar design is the development of the unitank. This single-tank process is designed to accomplish fermentation, aging, and conditioning in one vessel. Since no transfers are required, the chances for contamination are minimized, and cleaning and labor requirements are significantly reduced. These tanks must be designed with adequate cooling surface area, in zones located to provide for heat removal during fermentation, crash cooling, cold conditioning, and yeast cooling in the cone. Since some cellaring processes involve fermentation under top pressure to control the fermentation flavor profile (or provide natural carbonation), many of these tanks are designed to operate with significant overpressure, adding substantially to the construction cost. Unitanks designed for naturally carbonating ale by bunging may need to be designed for higher working pressures (up to 35 psi) than used in lager tanks, since the working temperature of the yeast is higher and the resultant equilibrium pressure is greater for the same carbonation specification. Since no transfer is performed after fermentation, dried foam clinging to the upper surfaces of the fermenter can be difficult to clean after prolonged aging. Although a cellar filled with unitanks may have a higher initial cost than some more conventional cellars, many new installations rely on this design because of the long-term savings resulting from the operational economies.

167. How are centrifuges used in the cellar?

Some modern breweries employ centrifuges to partially remove yeast (particularly flocculent strains) between primary and secondary fermentation and storage. These centrifuges are usually designed to remove only a percentage of the yeast in suspension, or a bypass loop is established to ensure that adequate yeast is carried forward to complete any subsequent maturation (see Chapter 2, Volume 2).

168. What is dry hopping?

Dry hopping is the practice of adding hops or, in some cases, hop products to the beer during aging in the cellar or in the keg or cask. Unboiled hops impart different hop qualities to a beer than the hops used in the standard process of kettle boiling. Very little bitterness is extracted since boiling is required to extract and isomerize the bitter resins. Traditionally, whole-leaf or shredded hops are used by placing them in mesh or cloth bags and submerging them in the cellar tank for a period of several days to weeks. Dry hopping is safest if carried out on end-fermented, moderate- or high-gravity beers since unboiled hops can be a potential source of contamination. Depending on the desired flavor intensity and oil content of the hop, rates can range up to 1 lb/bbl for some distinctive beers.

Brewers have used hop pellets for dry hopping, since they are much easier to deal with than filling and removing bags of whole hops. Hop pellets can be added directly to the tank and then removed during normal clarification. The fine grinding required for processing hop pellets tends to give them a less desirable flavor, with vegetative and coarser flavor qualities.

169. What are cask ale finings?

Brewers have long utilized methods to assist in the clarification of beer at various steps in the production process. Several materials and methods are utilized in cellar and cask beer production to assist in the settling and removal of yeast and protein.

The most widely utilized and effective clarification aid in the traditional cask ale cellar is isinglass, which is derived from the swim bladder of several species of tropical and subtropical fish. The active component in isinglass that is responsible for its clarification property is the protein collagen. The collagens derived from different fish species have slightly differing amino acid sequences and produce isinglass finings with varying properties and levels of effectiveness as clarifying agents. Care must be taken when preparing, handling, and storing the isinglass, since collagen is rapidly denatured by heat. The temperature should ideally be kept below 59°F (15°C) to maintain maximum effectiveness.

The clarification reaction is not totally understood, but the following probable mechanisms are the accepted theories:

a. The positively charged isinglass interacts with the negatively charged yeast cell to form a stable floc. The floc then precipitates and settles from the beer following Stokes' Law.

b. The soluble collagen forms a precipitate with soluble beer constituents and entraps the yeast and protein as it settles out, again following Stokes' Law.

The isinglass must be prepared by "cutting" (dissolving) in acid or must be purchased in an acidic standardized formulation. Isinglass preparations are sold based on their nitrogen content. Dose rates to the cask of 20–100 ppm (based on raw isinglass weight per volume) are typical, but actual process trials should be performed to quantify the effectiveness since varying grades and types will have a marked influence on the outcome. Solid isinglass preparations are rehydrated prior to use per the manufacturers' instructions and should be well dispersed and homogenously mixed with the beer to give even treatment. Cask beers are often fined in-line or dosed directly into the cask at racking. SO_2 is utilized as a preservative and stabilizer and will contribute some additional sulfur to the finished beer. Recent concerns with food allergies and seafood have prompted some countries to adopt or propose labeling requirements for beers containing isinglass.

Auxiliary finings are usually used in conjunction with isinglass and are composed of acidified silicates or acidic polysaccharides. These types of finings are highly negatively charged and bind with proteins, imparting a net negative charge that renders them available for isinglass interaction. In hard-to-clarify beers, this helps to create larger flocs, which settle more rapidly and completely.

Gelatin finings derived from animal skin and hoofs are also used as clarifying agents by brewers for cellar, but not cask, use, although they are not as effective as isinglass.

In any event, the amount of finings used is normally tested for each batch of beer to be racked. The flocculated sediments are expected to be stable enough to withstand the rigors of pub cellar delivery and to resettle to leave the beer brilliantly clear.

170. What other fermenting systems are in use today?

Worldwide, many different methods and fermentation systems are employed to produce a wide range of cereal beverages, all referred to as "beer." Historically, these brewing systems evolved based on available technology and equipment, local raw brewing materials, climate, brewing innovation, and product acceptance. Although many variations exist, the easiest classifications are based on the fermentation yeast type and the beer style produced. Breweries now produce a wide range of beer types utilizing a common fermentation system by varying raw materials and fermentation parameters. Most beer is now produced throughout the world utilizing bottom-fermentation methods with either modern large, conical fermenters or a variation of the classical lager method.

The following is a brief description of some unique fermentation systems in limited use today:

a. Union system. This very labor-intensive system is still being employed by at least one brewery in the United Kingdom. After the wort is inoculated with yeast and a short primary fermentation is carried out in rectangular flat-bottomed fermenters, the beer is transferred to a series of small wooden barrels (6 U.S. bbl each) that are fitted with swan necks that direct the yeast and fob to an open overhead collection trough. (Fob is the frothing or foaming of beer during processing or pouring.) Yeast can be harvested from this self-skimming design, and there are provisions to

Figure 1.54. Upper swan necks and collection troughs in the Union system. Yeast and beer from the fermentation mix and drain continuously back down to the casks below.

return the beer that is separated from the fob back to the barrels below. Modern installations have provisions for individually cooling each barrel. This system was an ingenious adaptation of the old-fashioned "cleansing" process, in which the fermenting barrels were allowed to foam over in order to expel excess yeast and trub. In the Union process, the brewer is able to reclaim this "foam over" beer while also harvesting a particularly powdery yeast strain for reuse. The small batch size, high labor and capital costs, and the hygiene challenges of wooden fermenters limit this method of production. *Figures 1.54* and *1.55* show the upper and lower parts of the Burton Union System, respectively.

   b. Shallow-pan or coolship fermentation.   Shallow-pan fermentation vessels are designed like coolships, which were historically used for wort cooling and trub separation. The vessels are designed with very shallow fill heights of around 12–24 in. (30–60 cm) and a flat, slightly sloping, pitched bottom and bottom side outlet to allow the removal of the liquid with little disruption of the precipitated hot-break sediments. Before whirlpools and closed coolers were in use, this design utilized the large surface area to promote atmospheric cooling and the shallow depth to decrease settling time for protein separation. These same principles al-

Figure 1.55. Casks (6 U.S. bbl) on the lower section of the Union system. Fermentations force beer and yeast up through tubes to upper collection troughs.

lowed for the control of fermentation temperatures before refrigeration was widely available and also helped in cold-break removal as well as yeast sedimentation and removal. Innovative early brewers probably adapted copper or iron coolships as a necessity to produce beers utilizing lager yeast in areas that lacked natural ice supplies for cooling. The early beers utilizing this method were probably minimally processed after fermentation and directly transferred to casks for conditioning and dispense. One famous West Coast brewery that still utilizes this method of fermentation has adapted the process to meet modern-day requirements for hygiene, product consistency, and shelf stability with modern materials and technology.

The volume of these fermentation tanks is naturally limited due to their shallow depth and required space. Forced CO_2 ventilation is mandatory, and depending on a brewery's ambient climate, the room may need to be supplied with conditioned air to achieve consistent fermentations year around. Because of the large footprint required, manual labor for cleaning, high construction costs, CO_2 collection and ventilation issues, and the added sanitation challenges, these installations are limited to the production of a few traditional beer styles.

c. Dropping or pumping-over systems. Ale breweries utilizing flocculent strains of yeast have developed methods to maintain the yeast in suspension to allow for dependable fermentation performance. One method involves transferring or dropping the fermentation from one vessel to another. This also allows the removal of dead yeast and sedimented cold break. Other methods involve pumping the wort from below over the top of the fermentation, helping to keep the yeast in suspension and adding some additional oxygen.

d. New technologies.   Many innovations have been experimented with in regard to fermentation systems over the years, such as continuous-fermenter designs (only one large-scale plant is still in operation), bioreactors, and immobilized yeast systems. These discussions are beyond the scope of this book.

REFERENCES

Atkinson, B., Brown, P., Frisby, R., Heron, P., Hudson, J., Laws, D., Lloyd, W., Putnam, R., Reed, R., Tann, P., Tub, R., and Jackson, G. 1985. *Manual of Good Practice for Cask Conditioned Beer.* Brewers' Society and Brewing Research International, Nutfield, England.

Bamforth, C. 2001. Beer flavour: Sulphur substances. *Brewers' Guardian* 130(10): 20–23.

Brandl, A., and Englmann, J. 2001. Trübung von Hefeweizenbieren. Technische Universität München, Wissenschaftzentrum Weihenstephan für Ernährung, Landnutzung und Umwelt.

Coghe, S., Martens, E., D'Hollander, H., Dirinck, P. J., and Delvaux, F. R. 2004. Sensory and instrumental flavour analysis of wort brewed with dark specialty malts. *Journal of the Institute of Brewing* 110:94–103.

Coghe, S., D'Hollander, H., Verachtert, H., and Delvaux, F. 2005. Impact of dark specialty malts on extract composition and wort fermentation. *Journal of the Institute of Brewing* 111:51–60.

de Piro, G. 2004. Beer flavor primer #1: Diacetyl. Online, www.evansale.com/diacetyl_article.html (C. H. Evans Brewing Company).

Delvaux, F., Depraetere, S. A., Delvaux, F. R., and Delcour, J. A. 2003. Ambiguous impact of wheat gluten proteins on the colloidal haze of wheat beers. *Journal of the American Society of Brewing Chemists* 61:63–68.

Delvaux, F., Combes, F. J., and Delvaux, F. R. 2004. The effect of wheat malting on the colloidal haze of white beers. *Master Brewers Association of the Americas Technical Quarterly* 41:27–32.

Derdelinckx, G., Vanderhasselt, B., Maudoux, M., and Dufour, J. P. 1992. Refermentation in bottles and kegs: A rigorous approach. *Brauwelt International* 2:156–163.

Dickel, T., Krottenthaler, M., and Back, W. 2002. Investigation into the influence of residual cold break on beer quality. *Brauwelt International,* pp. 23–25.

Hardwick, W. A., ed. 1995. *Handbook of Brewing.* Marcel Dekker, New York.

Herrmann, M., Back, W., Sacher, B., and Krottenthaler, M. 2003. Increasing beer ester levels through variations in wort composition and mashing procedures. *Monatsschrift für Brauwissenschaft* 56:99–103/106.

Heyse, K.-U. 1994. *Handbuch der Brauerei-Praxis.* Getränke-Facherlag Hans Carl, Nurnberg, Germany.

Kessler, H. 2002. Cold break removal from cold wort—Yes or no? *Brauwelt International,* pp. 76–79.

RECOMMENDED READING

American Society of Brewing Chemists. 2004. *Methods of Analysis.* 9th ed. ASBC, St. Paul, Minn.

Bamforth, Charles W. 2002. *Standards of Brewing.* Brewers Publications, Association of Brewers, Boulder, Colo.

Boulton, Chris, and Quain, David. 2001. *Brewing Yeast and Fermentation.* Blackwell Science, Oxford.

Holle, Stephen R. 2003. *A Handbook of Basic Brewing Calculations.* Master Brewers Association of the Americas, St. Paul, Minn.

Hough, J. S., Briggs, D. E., Stevens, R., and Young, T. W. 1982. *Malting and Brewing Science.* 2d ed. Chapman and Hall, London.

Kunze, Wolfgang. 1996. *Technology Brewing and Malting.* International ed. Versuchs- und Lehranstalt für Brauerei, Berlin.

Lewis, M. J., and Young, T. W. 2002. *Brewing.* 2d ed. Kluwer Academic, Plenum Publishers, New York.

McCabe, John T., ed. 1999. *The Practical Brewer: A Manual for the Brewing Industry.* 3d ed. Master Brewers Association of the Americas, Wauwatosa, Wisc.

Mitteleuropaeishe Brautechnische Analysenkommission (MEBAK). 2002. *Brautechnische Analysenmethoden.* Band II. Freising-Weihenstephan, Germany.

Smart, Katherine, ed. 2003. *Brewing Yeast and Fermentation Performance.* Blackwell Science, Oxford.

CHAPTER 2

Clarification, Filtration, and Finishing Operations

Grant E. Wood

Boston Beer Company

1. What is beer stabilization?

Stabilization is the process of preparing beer for serving or packaging. After stabilization, the initial fresh quality of beer is maintained over time. A brew pub may only have to ensure that beer is clear and fresh for one or two weeks, whereas a large brewery that exports to other countries may require up to 12 months of shelf life. Depending on the style, beer may have to be stabilized to varying degrees. The process may involve clarification, colloidal stabilization (the reduction of agents that cause chill haze), microbiological stabilization (the removal of yeast and spoilage bacteria), and carbonation adjustment.

2. What is shelf life?

Shelf life is the period during which a beer remains saleable and conforms to the brewer's accepted flavor profile. Beer ages from the time that it is brewed. Most brewers agree that there is a beneficial aging that occurs in the lagering tank. "Green beer" becomes smoother and more drinkable after a couple of weeks. However, after beer has been clarified, stabilized, and packaged for sale, its freshness begins to deteriorate. If steps are taken in the brewery to ensure sanitation, selection of good raw materials, good oxygen control, gentle treatment of the beer, and so forth, the flavor profile of the beer will be maintained for a longer time, and the beer will be as fresh as possible when it reaches the consumer.

3. Why should beer be stabilized?

Stabilization can enhance the clarity of beer, making it more pleasing to the eye. It can extend the shelf life of beer, and it can adjust the flavor and mouthfeel of some beers to enhance drinkability.

4. What is chill haze?

Chill haze is a nonpermanent haze in beer, formed by polypeptides (proteins) in combination with polyphenols or anthocyanogens (tannins). It is also referred to as a colloidal haze. A colloid is a stable mixture of two different forms of matter—in this case, a microscopic solid particle (protein-tannin complex) in a liquid (beer). Chill haze forms when a beer that is clear at room temperature is cooled to serving temperature. More broadly, colloidal hazes may form at higher temperatures and may have other, less obvious causes.

Chill haze tends to form in all-malt beers and in beers made from high-protein or poorly modified malts. Other factors promoting haze formation include

> poor kettle boil
> high storage or lagering temperature
> high levels of copper and iron
> agitation in the package during shipping
> high-temperature storage of packaged beer during shipping or
>   warehousing
> dry hopping
> high levels of dissolved oxygen
> poor initial filtration

Chill haze usually becomes permanent over time. The reaction that forms the haze is initially reversible, but over a matter of weeks it becomes a one-way reaction, as gradually larger particles form. The particles may eventually become large enough to precipitate and settle to the bottom of the package, forming a dust or a clumpy deposit, which then clouds the beer as it is poured into a glass.

5. How is chill haze measured?

The American Society of Brewing Chemists (*Methods of Analysis*, method Beer–26) measures turbidity in formazin turbidity units (FTU). Turbidity can be measured by comparing filtered beer to standards in a light box; with a nephelometer, a bench-top instrument; or by in-line measurement of haze. Turbidity-measuring instruments work by measuring either the transmission or the scatter of a light beam as it passes through a sample of beer. The measurement of turbidity can be affected by the color of the beer. The darker the beer, the more difficult it will be to measure the deflection of light by haze-causing particles. Several companies supply bench-top and in-line instruments for measuring haze. With automation, it is possible to monitor the outflow of a filter or centrifuge with a haze meter and send the filter into recirculation if turbidity readings are out of spec.

6. What other factors cause haze in finished beer?

Bacterial infections can cause haze in beer, with the bacteria themselves causing the haze. More rarely, haze is caused by starch that remains in beer from an incomplete conversion in the mash tun. Calcium oxalate (beer stone) can redissolve in beer under certain circumstances and form a haze. Some cleaning compounds, sanitizers, and coolant compounds can cause haze if they are allowed to get into the beer through a failure of quality control. An example is quaternary ammonium compounds, which are typically used for sanitizing non-beer-contact surfaces.

Microscopic examination of beer can reveal oxalate crystals, bacteria, or unconverted starch granules. A change in typical pH readings in a beer could indicate chemical contamination, but often a very small amount of a contaminant causes haze, and the immediate cause may be difficult to ascertain.

If there is any deviation from the normal clarity profile of a beer, it should be thoroughly investigated before the product is released for sale, to ensure the safety of the consumer and the reliability of the clarification regime.

7. What process aids can be used to aid in colloidal stability?

In the recent past, American brewers have used papain, an enzyme that breaks down proteins into smaller pieces called peptides, to reduce the effect of protein chill haze. Peptides are small chains of amino acids derived from the breakdown of malt proteins. The problem with papain is that it also breaks down peptides that enhance head retention, producing brilliant but flat-looking beers. When papain was used for enzyme stabilization, propylene glycol alginate (PGA) would also be added, to help restore the foam taken away by the papain.

Many brewers use silica gels to stabilize their beers. Silica gel, when hydrated properly, acts like a microscopic sponge to adsorb proteins from beer. Microscopic pores in the gel house electrically charged active sites that attach and bind to proteins of a certain size. Silica gels are very selective and have shown no evident effect on foam-assisting proteins. They are most effective when mixed in a well-agitated tank from a slurry containing approximately 1 lb of silica per gallon of deaerated water (see question 32). The silica is allowed to hydrate for a period of 30 minutes, while it is kept in solution by mixing the slurry. To help keep dissolved oxygen at a minimum, carbon dioxide is bubbled into the solution during the hydration period. The slurry is then proportioned into the beer as it is being filtered. Yeast tends to decrease the effectiveness of silica gel, so the gel is usually added after primary filtration or centrifugation and then removed during the polish filtration step. Typical dosage rates for all-malt beers are from 0.15 to 0.3 lb per barrel. The optimum amount will vary, depending on the beer, so trials should be performed to optimize the dosage. Silica gels need a minimum of 15 to 20 minutes of contact with the beer to optimize adsorption. Since most beer is treated after an initial clarification, a surge tank is used to allow sufficient time for contact between the beer and the gel.

Silica gels are supplied in two forms: hydrogels, which contain up to 70% water, and xerogels, which have been dehydrated to as little as 3% water. Xerogels contain more silica per pound than hydrogels and thus are used in lower dosages and have lower shipping costs. However, xerogels are very powdery, and handling them may cause dusting problems for employees. Some suppliers sell blends of hydrogels and xerogels, to gain the benefits of both.

8. How are tannins removed from beer?

Tannins can be removed from beer by treatment with polyvinylpolypyrrolidone (PVPP), an adsorbent, which acts on tannins much as silica gel acts on protein. PVPP particles are microscopic and have a porous, sponge-like structure to which tannins adhere. The particles then drop out of solution to the bottom of the surge tank or are trapped by a filter. PVPP handles like a xerogel; it is powdery and needs a hydration period to become effective. An optimal slurry (dosing) concentration is 1 lb per gallon of deaerated water. PVPP acts almost instantaneously, so there is no need for the lengthy contact time required by silica gel.

PVPP is available in different grades. Single-usage PVPP is filtered or washed out of the tank after settling. Regenerable PVPP, used by some large brewers in Canada and Europe, is collected in a separate filter system and regenerated by caustic. The PVPP is can be completely recovered, with very little loss. Regeneration may be too costly for small brewers, however, because of the initial capital outlay.

Overdosing with PVPP can affect the flavor and mouthfeel of beer. Polyphenolic compounds have an astringent and occasionally bitter taste. They also have some antioxidant effect. If too many polyphenolic compounds are removed, the beer may become less astringent and less bitter-tasting, and it may oxidize more quickly and thus have a reduced shelf life.

Stabilization agents are removed through filtration or settling. They do not remain in the beer, so they are not classified as additives or preservatives.

9. How is beer clarified?

Beer can be clarified by gravity over a long storage period or by the use of a fining agent, a centrifugal separator, or a filter.

10. Is there a way to clarify beer without mechanical means?

Beer can be clarified to some extent by gravity over the course of time or by the addition of a fining agent in the lagering period. Stokes' law (*Figure 2.1*) can be used to determine the velocity (v_g) of a particle settling in a liquid under the force of gravity (g):

$$v_g = [d^2(\rho_p - \rho_l)g]/18\eta,$$

where d is the diameter of the particle, ρ_p is the density of the particle, ρ_l is the density of the liquid, and η is the viscosity of the liquid.

Some ales are not intended to have long storage periods, and a fining agent is used to help yeast and protein flocculate and fall to the bottom of the tank, clarifying the beer more quickly. The fining agent increases the particle size by agglomeration and therefore increases the settling speed. However, beer will usually not be crystal clear even after some weeks of storage. Typical American malts have higher protein levels than their European counterparts. With the high protein levels and the heavy hopping rates of some popular beer styles, an insoluble protein-tannin chill haze can form, which will not settle out of the beer.

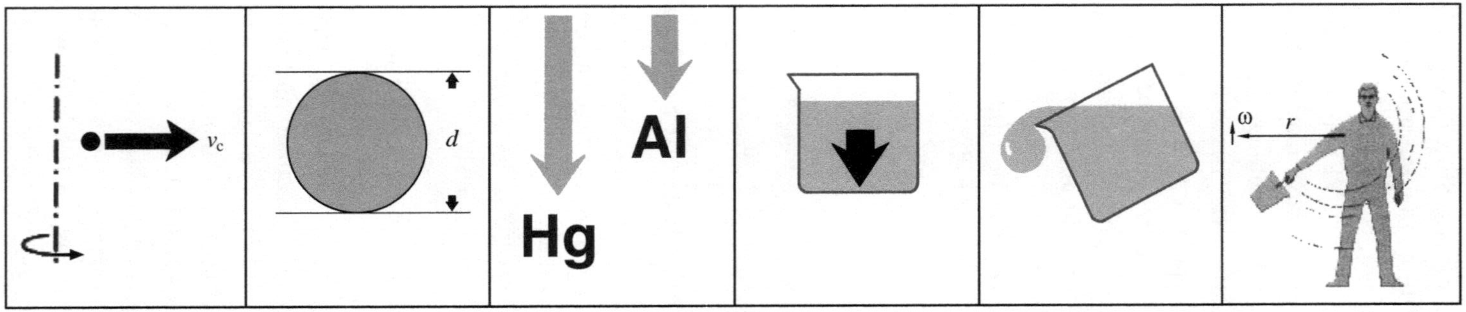

Figure 2.1. Stokes' law relates the settling velocity of a particle in a liquid to the particle diameter, particle density, liquid density, viscosity of the liquid, and centripetal acceleration of the liquid. (Courtesy of Alfa Laval)

Table 2.1. Fining agents

Mode of action	Activated carbon	Gelatin	PVPP[a]	Silica gel	Tannic acid	Isinglass
Color removal	×	×	. . .	. . .	. . .	. . .
Tannin reduction	. . .	×	×	. . .	. . .	. . .
Protein reduction	. . .	. . .	. . .	×	×	. . .
Yeast flocculation	. . .	×	. . .	. . .	. . .	×

[a] Polyvinylpolypyrrolidone.

11. What are finings?

Finings are agents added to beer to aid in clarification. Various kinds of finings are used, with different modes of action (*Table 2.1*). Some fining agents have an electrical charge, whereas others form chemical bonds with haze-causing compounds in beer. Some finings have adsorptive power and act like microscopic sponges to "soak up" haze-causing compounds. Some finings are used for specific problems, in which a target compound reacts with the surface of the fining agent. The target compound attaches to the fining agent, and larger particles are formed, which settle more quickly and completely to the bottom of the tank. For example, tannic acid has been used as fining agent and process aid. By the direct addition of tannin to beer, the colloidal reaction is tipped to favor the coagulation of the protein-tannin complex. The complex grows in size over time and settles out in the tank. This action is different from agglomeration with gelatin and adsorption by silica gel.

12. How are finings used?

Finings are typically added to "green," freshly fermented beer on the way to the storage or lagering tank. The beer is chilled in the tank or through a heat exchanger to as close to 30°F as possible. This forms a haze and increases the efficiency of the fining agent. The fining agent is usually mixed into a slurry in warm deaerated water or beer and allowed to rest and hydrate. The slurry is then added to the beer as it is being transferred from the fermentation tank to the storage tank. The best way to add finings is with an injection pump to proportion the slurry into the beer stream as the beer is being transferred, but the "bucket-through-the-door" method works as well. The finings, yeast, and protein will settle with gravity. Horizontal tanks work well for settling, by reducing the depth of the liquid and distance for particle settling and therefore reducing the time required for clarification.

Figure 2.2. Horizontal leaf filter. (Courtesy of Filtrox)

13. What kinds of filters are available for beer clarification?

The most common filters for primary beer clarification are diatomaceous earth (DE) or *kieselguhr* filters. Other types are sheet filters, pulp filters, and cartridge filters.

14. How do DE filters work?

In a DE filter, beer is filtered though a bed of diatomaceous earth held on a support, usually a stainless steel screen with a 60-micron mesh. The flow of beer through the support, also called the septum, lodges DE on the support in a fine bed that traps yeast, bacteria, and large particles of protein-tannin complex, to clarify the beer. Continuous flow through the filter bed is required, to make sure it stays in place on the support. If there is an interruption in the flow of beer, the DE can be displaced, and a loss of filter integrity occurs. Typical interruptions or shocks are caused by power interruptions, rapid opening or closing of valves, and deadheading the filter (closing the outlet and stopping the flow).

15. What types of DE filters are available?

Various types of DE filters are available: vertical leaf, horizontal leaf, plate-and-frame, and candle filters.

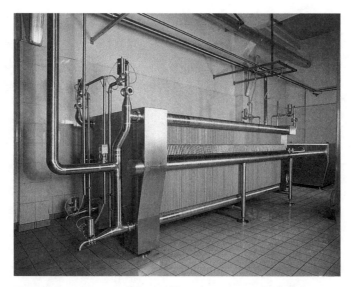

Figure 2.3. Plate-and-frame filter. (Courtesy of Filtrox)

a. Leaf filters.   Vertical and horizontal leaf filters (*Figure 2.2*) are the most typical in the brewing industry. They resemble a small tank. The tank contains a series of hollow screens (the leaves) mounted on a central pipe shaft. Beer flows into the tank, or shell, passes through the screen and into the hub, and proceeds out of the filter through the hollow central shaft. Leaf filters offer fast throughput, low operating labor cost, and fast turnaround but have a high capital cost.

The filter is usually accompanied by one or two small tubs or mixing containers in which DE is mixed in a slurry with deaerated water or beer. A precoat is deposited on the filter to prepare it for filtration. The deposition of the precoat depends on the design of the filter flow. In one design, the slurry is circulated through the tub and back into the filter. In another design, the slurry is pumped into the filter, and then the filter is circulated upon itself through the main pump. Once the water or beer is clear, DE has been lodged on the screen, and the filter is ready.

In a vertical leaf filter, beer is introduced into the filter stream gradually while the pump is running, to keep a constant flow across the face of the screen. If the flow is lost, the precoat powder can drop off the screen. In a horizontal leaf filter, the filter powder lies on top of the screen. If the beer flow is disrupted, the precoat layer is more likely to stay intact.

The typical precoat rate is 500 g of filter powder per square meter of filter area. Some operators recommend two precoats, the first with a

Figure 2.4. Candle filter. (Courtesy of Filtrox)

coarse grade of DE combined with perlite or cellulose floc and the second with a thinner coat of a finer powder, which is also added to the beer stream entering the filter inlet.

b. Plate-and-frame filter.   A plate-and-frame filter uses a cloth or paper "screen" supported by a metal or plastic frame (*Figure 2.3*). The filter is precoated with DE, and filtration occurs as in a leaf filter. The plate-and-frame filter has a slower throughput and requires more labor and time to flush out and reset than the leaf filter but has a lower capital cost.

c. Candle filter.   In a candle filter (*Figure 2.4*) the leaves are replaced by hollow shafts wound with wedge wire at a very narrow width. The hollow shafts, or candles, are mounted vertically on a top or bottom plate and precoated with DE. Beer flows into the shell, through the DE, into the candle, and then out through the mounting plate. Candle filters offer lower shear stress and flow velocities, but they require a beer-to-water interface to clear the volume out of the filter housing bell. Shear stress may break up yeast cells or proteins settled out during the aging cycle, allowing smaller particles into the finished beer. Changing from water to beer to water during filtration can cause overdilution of the beer if not monitored properly.

16. How does a DE filter operate?

In all types of DE filters, a stream of DE slurry is added to the beer stream going to the filter inlet, known as the body feed. Slow dosing of the body feed acts to add depth to the filter bed and allows for longer filter runs. By constantly creating more filter surface, it prevents yeast and protein solids from building up and blinding the filter, forming an impermeable layer over the DE precoat. The amount of body feed that can be administered to the filter is limited by several factors: the total area of the filter support, the distance between supports, the body feed rate, and the amount of solids fed to the filter. If filtration continues too long, bridging can occur between leaves or candles, pushing the supports apart and damaging the filter. Consult the manufacturer for guidelines for each particular filter. To make a slurry for dosing body feed, filter powder and water are mixed in a ratio of 1:10 by weight (e.g., 1 kg of powder to 10 kg of water). During filtration, if there is a heavy load of yeast and other solids, more body feed may be required to prevent the filter from blinding. De-aerated water or, less preferably, beer should be used to make up slurries for precoat and body feed. Mixing beer in a precoat tub can cause excessive foaming and can oxidize the beer.

17. What conditions are monitored in DE filtration?

The filter operator should monitor three main conditions during filtration: temperature, filter flow rate, and differential pressure.

a. Temperature.   Beer should be kept at cellaring temperature (32–35°F) during filtration, to prevent redissolution of the cold-induced complex of proteins and tannins that causes chill haze. The low temperature also helps to keep the CO_2 in solution and discourages microbiological growth.

b. Flow rate.   A constant flow through the filter should be maintained to prevent shock to the filter bed and to allow even deposition of the body feed.

c. Pressure.   The differential pressure across the filter is the difference between the inlet pressure and the outlet pressure. Most filters cannot be operated at a differential pressure higher than 50 psi. Operation beyond this point can damage the filter support, damage seals and gaskets, or cause failure of the filter shell. All filters should be fitted with overpressure relief valves for safety. Consult the filter manufacturer's operations manual for a particular filter to find its safe pressure limits.

When the filter flow rate begins to fall, the differential pressure becomes too high, or the turbidity of the beer from the filter outlet becomes unacceptable, it is time to wash out the filter. The beer is then pushed out of the filter under CO_2, and the screens or candles are sluiced off with water jets. In small filters, the shell is opened, and the screens are rinsed manually. In large filters, the stack of leaves rotates while fixed sprays of water knock powder off the screens and out a large drain at the bottom. Spent DE can be collected and either recycled in various ways or discarded as landfill.

18. How do you know if you are using enough filter powder?

If possible, inspect the filter cake after a run, before the filter is sluiced and cleaned. Take a sample of the filter bed and rub it between your fingers. The sample should feel moist, not dry, but not slippery. If it is slippery or slimy, then you are probably not using enough powder. The yeast and protein are overloading the filter powder, and more should be added to the body feed. If the sample is too dry or gritty, you are probably using too much body feed and could cut back on the next run.

The initial turbidity readings of the beer to be filtered also provide information on how heavily the beer should be dosed. Turbidity readings throughout the run will allow the operator to monitor the effectiveness of the filtration.

19. What is diatomaceous earth?

Diatomaceous earth, or *kieselguhr,* is the fossilized remains of microscopic plants deposited on the floors of lakes and oceans millions of years

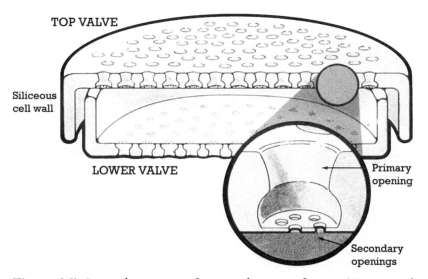

Figure 2.5. Internal structure of a typical marine diatom. (Courtesy of World Minerals)

ago. *Figure 2.5* shows the structure of a typical diatom particle. There are two general types of DE: marine and freshwater. The main difference between them is that marine DE has a wider variety of shapes and pore sizes than freshwater DE. Marine DEs (*Figure 2.6*) also contain a wider range of minerals that can dissolve in beer during filtration, particularly iron. Both types can do a fine job of filtration.

Diatomite deposits are mined, crushed, milled, and kilned at 800°C. Kilning removes organic material trapped in the DE. A flux, or binding agent, such as sodium carbonate, may be added during kilning to bind small particles of DE together, because most diatoms are too small to be effective in filtration. The DE is then sieved and sorted into various grades for use in filtration and other industries.

DE filtration is effective in the stabilization and preparation of beer for the consumer, but it does have some drawbacks. DE is a solid material, and after filtration is complete, the filter must be washed out. Used DE and yeast can place a high biological oxygen demand and total soluble solids load on the local sewer system, which may increase sewer surcharges. Care should be taken when deciding which type of filter system to use. Depending upon local codes, it may be necessary to install a settling tank to catch used filter material for disposal in a landfill. Diatomaceous earth may affect the quality of the beer. Because of the adsorptive capacities of DE, silica gel, and PVPP, all filter powders and filter aids should be

Figure 2.6. Microphotograph of marine diatomaceous earth particles. (Courtesy of World Minerals)

kept away from moisture and odors. In addition, some filter powders have a high soluble iron content, which can transfer a metallic flavor to beer during filtration. Most DE suppliers can provide low-iron products.

20. What is perlite?

Perlite is a naturally occurring volcanic rock (*Figure 2.7*). It is processed from its raw state by rapid heating to 870°C, which causes the rock to expand like popcorn. This expansion is caused by a small amount of water trapped within the structure of the rock. The vaporization of the water forms a fine microcellular structure, like that of a tiny sponge.

Perlite is available in various forms: powder, sheets for use in sheet filters, and cartridges. It can be used with or instead of DE for filtration in most DE filters. Perlite is not considered as effective a filter aid as DE, because of its smooth surface and lack of internal porosity. It may still make a good prefilter for the removal of most solids from storage beer.

21. Is exposure to diatomaceous earth or perlite filter powder harmful?

The International Agency of Research on Cancer has classified "inhaled amorphous silica from occupational sources" in Group 1 as a substance "carcinogenic to humans." The U.S. Occupational Safety and Health

Figure 2.7. Microphotograph of perlite. (Courtesy of World Minerals)

Administration (OSHA) has not classified amorphous silica as a carcinogen. However, chronic and prolonged exposure to silica dust can cause silicosis, a progressive and sometimes fatal lung disease. OSHA monitors the exposure of workers to silica dust. When handling any filter aid, workers should use proper personal safety equipment, as listed on the Material Safety Data Sheet (MSDS) of the product, and the work area should be properly ventilated.

22. What is floc?

Floc, also called filter aid, is a fibrous material used to help support filter beds (*Figure 2.8*). It is usually a cotton fiber called linters or a paper fiber. Floc is sometimes referred to as cellulose, which is the major constituent of pulp filters. Pulp filters are rarely used today. One benefit of these filters is that the pulp can be washed and reused for filtration, although this operation is very labor intensive.

Floc can be used by itself as a coarse filter, but it is more commonly used together with DE or perlite. It can be added to a precoat to strengthen the precoat bed with its fibrous structure. Floc is also used to

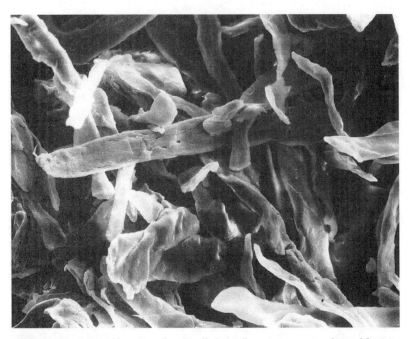

Figure 2.8. Microphotograph of cellulose floc. (Courtesy of World Minerals)

help cover small defects in filter screens or slightly damaged areas. The fibers bridge gaps in the screen and help the DE to build a matrix over minor imperfections.

23. What is a membrane filter?

Membrane filters have a fixed element, usually a thin polymer plastic with microscopic perforations or a woven screen, that filters out yeast, bacteria, and haze-causing particles. The membrane is often pleated, to increase its surface area. It is encased in a plastic cartridge that is inserted into a filter housing. Membrane filters are available from many manufacturers, in various materials and in a range of pore sizes. Typical membrane materials are polytetrafluoroethylene (PTFE) and polyethylene. The core membrane may be surrounded by an outer layer of wound polyethylene or nylon fiber to add depth.

These filters are expensive, and they tend to foul easily under medium to heavy loads of solids. For this reason they may not be suitable for primary or even secondary clarification. Membrane cartridge filters are often used as trap filters downstream of the main DE filters, to catch any bleed-through of DE particles or yeast. Membrane filters are also effective for sterile filtration, or microbiological stabilization of beer after it has been filtered by DE. A two-stage membrane consisting of a 0.65-micron prefilter and a 0.45-micron absolute filter can remove 99.9% of beer spoilage organisms from the product. Vigorous attention to sanitation downstream of this filter is essential to maintaining the "sterility" of the beer.

Recent improvements in the manufacture of membrane filters have greatly improved their operational life, and many can be effectively regenerated by back-flushing or cleaning in an approved solution. There are numerous manufacturers of membrane filters, using a wide range of materials. Exercise care when selecting a filter system. The very fine filtration ability of filtration media may remove some flavor-active compounds from the beer, "stripping" the flavor. Some materials may not be suited for regeneration or may deteriorate with excessive heat. Sterile membrane filtration should not be regarded as means of removing gross contamination from beer but can be a viable alternative to pasteurization to improve the stability of the product.

24. What is a sheet filter?

A sheet filter consists of preformed sheets made from a mixture of fiber and DE, fitted into a plate and frame housing and pressed together. Sheet filters can be effective for final filtration. Some can be back-flushed and cleaned for a longer life. They are not suitable for high-solids filtration, as they do not have the depth capacity that is supplied by a DE leaf filter with a body feed stream of DE.

Sheet filters are available in various sizes and in a range of filter ratings. Some are impregnated with PVPP or activated carbon. Some are fine enough to approach membrane filtration for microbiological stability.

The sheet filter may be the easiest type of filter to operate, but it has some drawbacks in comparison with a DE filter. It is more labor intensive to break down, clean, and resheet a sheet filter than to clean a DE filter. The pads may need to be wetted before the frame is closed. For cleaning, the filter sheets must be removed, and the plate and frame housing closed, to circulate a cleaning solution.

25. What is a trap filter?

A trap filter is a sock-, bag-, or cartridge-type filter installed at the final discharge of the polish filter or on the beer line that feeds a bottle filler or racking line. It is used to ensure that no DE or filter aids escape previous filtration and remain in beer destined for packaging. Trap filters are typically rated to catch particles 1 to 5 microns (µm) in diameter, the size of most filter aid particles.

26. What is a centrifugal separator?

A centrifugal separator, or centrifuge, is a device that accelerates the settling of particulates and colloidal matter from beer.

The oldest and most economical means of removing yeast and other particulates from beer is to hold the beer in a tank, allowing solids to settle under the force of gravity. Reducing the depth of the tank decreases the settling time. Settling plates in the tank further reduce the settling distance and therefore the time required for settling; however, adding settling plates to a modern sanitary brewing tank is not recommended.

The force of gravity cannot be increased, but centripetal force can be applied to a liquid by pumping it through a rapidly spinning chamber. The force of centripetal acceleration of a spinning chamber, replacing the force of gravity in Stokes' law (see question 10), greatly increases the settling velocity. The centripetal force at the perimeter of a centrifuge bowl can reach 4,000 times the force of gravity. The modern centrifugal separator, combining a centrifuge bowl and a series of settling plates, is very effective in clarifying beer (*Figure 2.9*).

In the brewery, the goal of centrifugation is the removal of the solid phase (yeast and other solids) from the liquid phase (beer or wort). The typical process centrifuge used in modern breweries is a solids-ejecting type, which is effective in removing 99.9% of the solids from beer. In the cellar, centrifuges have been used to remove yeast from beer being transferred from fermentation to storage and to collect yeast for repitching. They are also used after storage to reduce the solids load going to the filter. Clarification by centrifugation after storage can reduce the amount of filter powder required and the associated costs of usage and disposal.

Centrifuges are also used to "dial in" a specific amount of yeast to be left remaining in some beers, such as hefeweizens. By controlling the flow rate of beer through the centrifuge, the brewer can control the amount of yeast being removed and not filter out desirable colloidal haze.

Centrifuges require high capital outlay and are expensive to maintain. A centrifuge may increase the throughput of a brewery operation by reducing tank residence time in the fermenters, but this benefit must be compared to its expense. Improperly designed units may contribute to high levels of dissolved oxygen in the beer and impose shear forces that actually increase haze by further shattering colloidal haze particles.

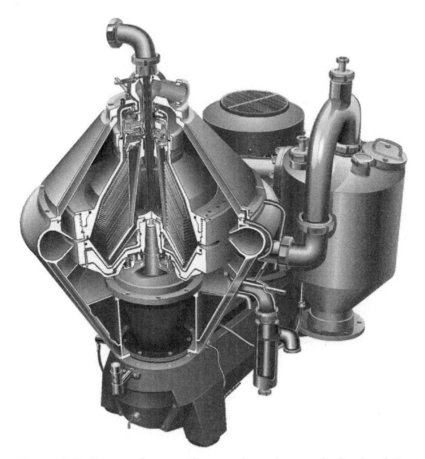

Figure 2.9. Cutaway diagram of a centrifuge, showing the bowl and the plate stack assembly. (Courtesy of Alfa Laval)

27. How should I clean my filter, and how often should it be cleaned?

Always refer to the manufacturer's operations manual to determine the cleaning regime for the filter you are using.

A good rule of thumb is to completely clean a filter if it has been out of operation for more than 24 hours. A solution of 1–2% sodium hydroxide at 160°F works well for cleaning filters. The filter shell, lines, pumps, and tubs should be completely filled. The filter should be cleaned at a speed higher than its normal operating speed, to apply additional force to soil that may be present inside the filter. It is important to flush deadheads in the filter with the cleaning solution (a deadhead is the dead end at the end of a pipe, such as the overpressure safety valve).

Sanitary steam can be used to clean some filters. *Check the operations manual!* For other filters and for long-term storage, iodophors, peracetic acid, chlorine dioxide, and other sanitizers are used (see Chapter 2, Volume 3).

It is a good idea to run a cleaning cycle with an acid cleaner quarterly or semiannually. A 2–3% solution of food-grade phosphoric acid (1 gallon of 85% phosphoric acid per barrel of water) will work well to keep oxalate beer stone from building up inside the filter.

28. What is dissolved oxygen?

All gases can dissolve into solution in a liquid (such as beer) to some extent. Carbon dioxide easily dissolves in beer to give it its characteristic fizz and foamy head. Oxygen can also dissolve in beer, and dissolved oxygen (DO) is the number-one cause of degradation of beer flavor stability. Excessive oxygen in beer causes a series of chemical reactions resulting in flavors usually described as "wet cardboard" or "papery."

29. What are some causes of high DO?

Oxygen uptake in the brewery can have various causes, of which the following are a few examples:

    a. poor connections between a tank and hose fittings, including loose Tri-Clamp fittings, especially on pump inlet connections

    b. transfer of beer through previously empty hoses exposed to the atmosphere

    c. filling a freshly washed, unpurged tank

    d. leaking pump seals

    e. leaking zwickles on transfer lines

    f. excessive exposure of beer to "city water" (bad interface of water to beer in piping)

    g. leaking tank doors

    h. fountaining, or splashing in the tank at the beginning of a transfer of beer from one tank to another

    i. unpurged filter powder in mixing and dosing tubs

A good rule of thumb is that anywhere beer can be seen leaking, oxygen is getting in. Since there is pressure in the line, tank, or hose, it may seem logical that oxygen is not getting in if beer is visibly coming out. However, because of flow across the gap, air will be dissolved in the beer. This occurs especially at leaks on the inlet side of pumps. The effect of oxygen uptake can be greatly reduced if carbon dioxide is used to purge mixing tubs of filter aids and DE. Deaerated water or CO_2 should always be used to purge lines and hoses before and after a beer transfer.

30. How is DO measured?

Meters measuring dissolved oxygen in beer are available from several manufacturers. The meter is attached to a sample point, usually a zwickle. Beer flowing through the meter passes across a sensor covered with a gas-permeable membrane. Oxygen passes through the membrane and causes a change in the conductivity of a gel surrounding an electrode. The change in conductivity provides a measure of the amount of oxygen dissolved in the beer, in parts per million (ppm) or parts per billion (ppb). Desirable readings are below 150 ppb—the lower the better, with the target being 0.

31. What are some ways of reducing DO?

Good beer-handling practices in the brewery will help alleviate oxygen uptake. The following practices can reduce the amount of DO in beer.

a. Tank purging.   After a storage tank or bright beer tank has been washed, it should be purged. Purging can be done in two ways: by gassing the tank from the bottom with carbon dioxide or by water gassing.

As carbon dioxide is heavier than air, gassing the tank from the bottom with carbon dioxide pushes air up and out the vent. After purging for a period of time, the tank can be pressurized with carbon dioxide to help prevent foaming and release of gas during transfer. A blanket of carbon dioxide over the beer in the tank will also help protect the beer.

Water gassing is an alternative method of tank purging. The tank is completely filled with water, which is then pressed out with CO_2, to eliminate all air. The water can be pressed into a neighboring tank and reused several times for purging before finally being pressed to the sewers.

b. Line purging.   Lines and hoses between tanks should be equipped with valve tees to allow the lines to be purged with gas or filled with deaerated water. When beer is to be transferred, the supply tank is opened, the water or gas is flushed out to the floor until beer is at the tank, the tee is shut, and the receiving tank is then opened.

c. Careful transfer of beer.   Gentle transfer of beer from tank to tank is important. Splashing or excessive foaming will cause the beer to absorb oxygen from the tank atmosphere more readily. Variable-speed drives for beer transfer pumps can control the speed of transfer to reduce splashing, or fountaining, in the tank and reduce mechanical shearing of the beer. The transfer should be started slowly until there is enough beer in the tank to prevent fountaining. The speed can then be increased to fill the tank. Before the end of the transfer, the flow should be slowed to make it easier to empty the supply tank and to prevent vortexing. The beer can also be throttled at the receiving tank, but this may cause more turbulence coming into the tank. A sight glass is useful on the inlet side of the pump to indicate visually when the tank is empty. Alternatively, sensors are available that will signal or shut down the pump when a line goes empty. It is important to monitor the end of the transfer to avoid pulling air into the line. The line can then be "pushed out" with water, preferably deaerated water, or CO_2.

d. CO_2 dusting. Forcing CO_2 into beer while it is being transferred helps keep DO at a minimum. If the DO in a tank is high after transfer, and the tank has carbonating stones, oxygen can be purged from the tank by bubbling CO_2 through the beer. However, compounds that enhance aroma and foam head retention may also be removed with the DO.

32. What is deaerated water?

Deaerated (DA) water is potable brewing water that has been treated to remove as much dissolved oxygen as possible. Dissolved oxygen can be removed from water by machines made specifically for that purpose. These machines heat the water and then spray it across a packed column of ceramic saddles to increase the surface area. Vacuum is applied to the column to reduce the internal pressure and encourage the oxygen to leave the water. The water is then chilled and force-carbonated. DA water is good for purging packing lines before and after a transfer of beer and packing filters, to reduce dissolved oxygen in the beer.

A low-tech method of removing dissolved oxygen from water is to boil it for an hour to remove chlorine and dissolved oxygen in the kettle, then cool the water, transfer it to a tank fitted with carbonation stones, and purge it with CO_2 through the stones until the DO content is below 100 ppb. DA water can also be made more slowly by simply purging a tank of dechlorinated water with CO_2 through carbonation stones. This method is time-consuming and possibly more expensive than boiling, depending on the cost and availability of CO_2. The water should be kept at cellar temperatures, and the tank should be cleaned regularly.

33. What is potassium metabisulfite (KMS)?

Potassium metabisulfite ($K_2S_2O_5$), or KMS, is a salt commonly used in wines (in concentrations of up to 30 ppm) and also used as an antioxidant in beer. In the United States, KMS added to beer in concentrations of less than 10 ppm is not required to be declared on the label. This compound gives a slight measure of protection against the development of oxidized flavors. The activity of KMS is not exactly known. One view is that it acts as an oxygen scavenger, soaking up excess oxygen dissolved in beer. Another possibility is that it reacts with the carbonyl compounds that cause papery and aged flavors in beer and renders them tasteless.

Care must be taken in the use of KMS. Beer naturally contains sulfates. If beer is brewed with a yeast that produces a high level of sulfur compounds, adding more sulfur in the form of KMS could put the sulfur content over the legal limit for the beer (level at which labeling is required). In addition, some people are sensitive to sulfites and can have adverse allergic reactions if exposed to even small amounts.

KMS can be used either in the transfer from fermentation to storage or in the final filtration of beer. It can be purchased in its dry form and then dissolved in water for addition to beer. When KMS dissolves, sulfur dioxide (SO_2), a pungent, suffocating gas, is released. Use a respirator in an adequately ventilated area when dissolving the powder. The dissolution is an endothermic reaction, which cools the solution as the compound dissolves.

34. What is carbonation?

During fermentation, yeast digests the sugars produced in the brewhouse and creates ethanol, carbon dioxide, and other flavor products. Carbon dioxide produced during fermentation dissolves in the beer. This dissolved carbon dioxide is called carbonation. Carbon dioxide dissolves readily in water, because of the slightly negative electrical charge of the CO_2 molecule. It will grab onto water molecules and form a weak acid, carbonic acid. During unpressurized fermentation, 1.2 to 1.7 volumes of CO_2 per volume of beer are left in the beer at atmospheric pressure. Normal serving levels of carbonation in beers are between 2.3 and 2.8 volumes. Carbon dioxide dissolves more easily as the temperature is reduced and the pressure increased. Procedures for adjusting the carbonation of beer are usually performed at low temperatures in pressure-rated tanks.

35. How is carbonation measured?

Carbonation is measured in volumes of CO_2 per volume of beer. Two devices are used to measure carbonation: the Zahm meter and the Gehaltemeter.

The Zahm meter has a port that attaches to the tank, an inlet valve, a thermometer, a pressure gauge, and an outlet valve. The meter is attached to the sample port of the tank, and beer is allowed to flow through it until the bottle is the same temperature as the beer. The valves are then closed, and the bottle is removed from the tank and shaken vigorously. Temperature and pressure readings are then checked against a chart that calculates the volume of CO_2 in the beer (see Chapter 5, Volume 2).

The Gehaltemeter also attaches to the tank, and a sample of beer is allowed to flow through the device. When the beer has run through and eliminated any foam from the sample chamber, the valves are shut. An electrode in the bottom of the sample chamber releases the gas from the sample. Sensors in the chamber measure pressure and temperature. With this information the device calculates the volume of CO_2 in the beer.

36. How can carbonation be adjusted?

The brewer can choose not to carbonate through the use of counterpressure fermentation or kräusening. Beer can be mechanically carbonated in-line during a transfer or in-tank after filtration.

a. In-line carbonation. Clean, dehydrated carbon dioxide is forced into the beer through a porous stainless steel plate while the beer is flowing to a pressurized receiving tank. Care must be taken to maintain a tank counterpressure between 12 and 15 psi while filling, to keep the CO_2 in the beer and to minimize foaming.

b. In-tank carbonation. CO_2 is forced through one or more carbonating stones, made of porous stainless steel or ceramic, in the bottom of the tank. The porous carbonating stones produce fine bubbles and facilitate the dissolution of carbon dioxide in the beer. For faster carbonation, some pressure is bled off the top of the tank while CO_2 is forced in the bottom. This method can cause some loss of beer volatile aromas and a reduction in head-forming potential. The final tank pressure will depend on the temperature of the beer. The warmer the beer, the higher the pressure needed to keep the CO_2 in solution. For forced carbonation, the gas pressure of the CO_2 going in should be approximately 40 psi. Raising the CO_2 level in beer by one volume takes 0.5 lb of CO_2 per barrel of beer.

The brewer should use only pressure-rated vessels for carbonation adjustments and for holding bright beer for packaging. The tank should have a certification stamp from the American Society of Mechanical Engineers (ASME), certifying its ability to withstand normal operating pressures. The maximum operating pressure of a bright beer tank is typically 20 to 25 psi. Bright beer tanks are usually protected from overpressure by a pressure relief valve. Various styles of pressure relief valves are available, including breaker bar–diaphragm, spring-loaded, and counterweighed valves. These are all set to a specific maximum pressure to protect the tank from rupturing.

REFERENCES

Alfa Laval Corporation. 1993. *Basic Separation Class High Speed Disk Centrifuges.* Alfa Laval, Warminster, Pa.

American Society of Brewing Chemists. 2004. *Methods of Analysis.* 9th ed. ASBC, St. Paul, Minn.

Bushnell, Sarah E., Guinard, Jean-Xavier, and Bamforth, Charles W. 2003. Effects of sulfur dioxide and polyvinylpolypyrrolidone on the flavor stability of beer as measured by sensory and chemical analysis. *Journal of the American Society of Brewing Chemists* 61: 133–141.

De Clerck, Jean. 1957. *A Textbook of Brewing.* Vols. 1 and 2. Transl. by Kathleen Barton-Wright. (1994 reprint.) Siebel Institute of Technology, Chicago.

European Brewery Convention. 1999. *Beer Filtration, Stabilisation, and Sterilisation Manual of Good Practice.* Fachverlag Hans Carl, Nürnberg, Germany.

Hardwick, William A., ed. 1995. *Handbook of Brewing.* Marcel Dekker, New York.

Leeder, G. 1993. Features of a modern filter line. *The Brewer* 79:948.

McCabe, John T., ed. 1999. *The Practical Brewer: A Manual for the Brewing Industry.* 3d ed. Master Brewers Association of the Americas, Wauwatosa, Wisc.

Niemsch, Klaus. 2000. Beer stabilization, going to the next millennium, with the experience of yesterday. *Technical Quarterly of the Master Brewers Association of the Americas* 37:455–458.

Noordman, T. Reinoud, Peet, C., Inverson, W., Broens, L., and van Hoof, S. 2001. Cross flow filtration for clarification of lager beer—Economic reality. *Technical Quarterly of the Master Brewers Association of the Americas* 38:207–210.

Oechsle, Dietmar, Ascher, R., and Feifel, K. 2000. A new stainless steel membrane support for horizontal pressure leaf filters. *Technical Quarterly of the Master Brewers Association of the Americas* 37:377–381.

Reed, R. J. R. 1986. Centenary review article beer filtration. *Journal of the Institute of Brewing* 82(2).

Reed, R. J. R., and Freeman, G. J. 1993. Beer filtration, fitting the filter-aid to the beer. *Ferment* 6(1).

Rehmanji, M., Gopal, C., and Molas, A. 2002. A novel stabilization of beer with Polyclar® Brewbrite™. *Technical Quarterly of the Master Brewers Association of the Americas* 39:24–28.

Ringo, Stefanie M. 2002. International Society of Beverage Technologists (ISBT) carbon dioxide guidelines. *Technical Quarterly of the Master Brewers Association of the Americas* 39:32–35.

Zoecklein, Bruce W., Fugelsang, Kenneth C., Gump, Barry H., and Nury, Fred S. 1995. *Wine Production and Analysis.* Chapman and Hall, London.

INTERNET SOURCES

www.alfalaval.com
www.basf.com
www.begerow.de
www.haffmans.nl
www.perlite.org
www.worldminerals.com
www.zahmnagel.com

CHAPTER 3

Racking Room Operations

Bob August

Majestic Packaging Solutions

1. What is racking?

The term *racking* "is applied to the process of transferring beer from storage tanks to transport containers" (Vogel et al., 1946, *The Practical Brewer*, p. 141).

2. What is the general procedure in racking room operations?

The objective of racking operations is to transfer finished beer or ale into kegs or casks, often of various capacities, while maintaining the utmost integrity and quality of the final packaged product. *Figure 3.1* shows full Sankey kegs exiting a racking line after being cleaned and filled in a series of steps described later in this chapter.

Of primary concern is a clean and sanitary workplace. Since draft beer is not pasteurized in the package and is seldom flash-pasteurized or sterile-filtered, cleanliness should be the brewer's highest priority.

Beer should be properly finished for its style and within specification before it is released for racking. If the beer is fully carbonated, it should be racked as cold as possible without freezing. If carbonation is to occur in the keg or cask, it should be racked at the appropriate cellar temperature.

3. What are some general considerations for establishing a racking operation?

The finished beer tank should be located near the filling equipment, if possible. If the tank is placed sufficiently high above the filling equipment, gravity delivery may be possible. If the tank is below the racking

Figure 3.1. Full kegs exiting a racking line.

line, or if the tank outlet is below the filling equipment, a pump should be used. The pump should be specified to deliver the product at a rate marginally above the maximum requirement of the line and at a pressure adequate to maintain carbonation (at least 15 to 20 psig above the equilibrium pressure of the product).

If beer is supplied directly to the racker without a pump, the tank may have to be pressurized to 15 or 20 psi higher than the beer equilibrium in order to get the racker to provide calm filling. Therefore, if the beer equilibrium pressure is 10 psi, the tank must be built to withstand 30 psi. If a pump is used between the tank and the racker, the tank could be built to withstand 15 psi and the pump could supply the additional pressure required. The racker would see the higher process pressure but not the tank. A tank built to 15 psi is considerably less expensive than a tank built to 30 psi.

If the finished beer cellar is some distance from the racking line, a small buffer tank can be included. The buffer tank should be constructed to the same standards as the finished beer tank: it should be sanitary in design and capable of meeting the pressure requirements of the racking operation. A good installation requires additional pumps, level indication, and usually computerized flow controls, introducing a level of sophistication that small breweries may wish to avoid.

Accommodations should be made for storage of the product, once packaged. The storage temperature should be maintained at or near the filling temperature. Warmer storage is not recommended. If the product is carbonating in the keg or cask, it may be moved to cooler storage once it has matured, to maintain freshness.

4. What U.S. federal regulations govern draft beer?

Federal, state, and local regulations governing draft beer should always be thoroughly investigated by the brewer before a product is released to the public. Some general guidelines to federal regulations are presented below. For complete information, the reader is advised to consult the U.S. government publications listed at the end of the chapter.

A keg must be permanently labeled with the brewer's name. The place of production must be permanently marked on the keg, bung, or tap cover or on a label securely affixed to the keg. For breweries with multiple production facilities, the place of production can be part of a list of all locations, as long as it is not given less emphasis than any of the others. If more than one location is shown on the keg, the place of production must be indicated by printing, coding, or other markings on the bung or tap cover or on a separate label. The coding system must allow a government official to determine where the keg was produced. The code must give a street address if the brewery has more than one operation in the same city. The regional compliance director must be notified before a coding system is used. A label or tap cover used for labeling must have a Certificate of Label Approval (COLA).

Alcoholic malt beverages must be marked, branded, or labeled with the following information:

> brand name
> class and type designation
> name and address of brewer (may be stamped or branded on the
>   container)
> country of origin, if the product is imported
> name and address of the importer, if applicable
> name and address of the packer, for malt beverages packed for
>   the holder of a permit or a retailer
> net contents (stamped or branded on the container)
> alcohol content, if required by state law
> health warning statement

No statement of payment of internal revenue taxes may be shown.

5. What are the advantages and disadvantages of the various types of cooperage?

There are two basic styles of kegs: open systems and closed systems. Each has its advantages for some brewers. Kegs of both styles are also made of various different materials, the particular advantages and disadvantages of which are described below.

Keg style

a. Open systems.   Open kegs are characterized by their barrel shape and open filling system, in which the keg is filled through a hole in its side. Once a keg is filled, a wooden or plastic plug called a bung is hammered or driven into the opening, sealing the keg. The bung must be removed for cleaning.

There are essentially only two remaining dispensing systems for open kegs in the United States: the Hoff-Stevens design and the Golden Gate design. These designs do not require automated cleaning or filling systems, and the only automation commonly available for these kegs is for cleaning. Manual inspection after cleaning is still necessary.

The open system is suitable for the addition of hops to the keg (dry hopping) and the addition of finings to cask-conditioned products.

b. Closed systems.   Closed kegs are easily identified by their typically straight sides and handled upper chimes. They have a single valve incorporating two ports, one for the product and one for dispensing gas. Closed kegs lend themselves to fully automated lines and labor savings. Unlike open kegs, they can be filled so that the finished package contains only a very low level of dissolved oxygen. Closed kegs can be cleaned and filled manually, but manual operations are not recommended if the full benefits of these kegs are to be received. Unlike open kegs, closed kegs are not easy to inspect.

Composition

a. Wood.   Wooden kegs have all but disappeared in the United States, although some brewers have begun using bourbon and wine barrels as aging vessels. Wooden kegs add complexity to the flavor of the finished product, and they provide novelty in the presentation of the product. They do not require automated filling or cleaning systems. However, wooden kegs are the most difficult system to maintain. Because of their rarity, it is no longer easy to repair them, except where a skilled coopersmith is available. Wooden barrels are heavier than metal kegs of like size. They are not suited to high-pressure delivery systems. They can require unusual tapping equipment not likely to be recognized in the trade. The uniformity of the volume of barrels can be questionable. Thorough cleaning and sanitation are impossible. Wood can shrink unless kept damp during storage. Dried-out wooden kegs can leak. The interior surface of the keg must periodically be pitched, or coated with a mixture of tree resin and cottonseed oil mixed with paraffin. Wooden kegs can provide a satisfying experience, but they are best suited to an environment in which the brewer can maintain complete control over the vessel.

b. Aluminum.   One of the first materials used for making metal kegs was aluminum. Its primary advantage is its light weight. Aluminum can be shaped into any modern configuration. However, it is very easy to damage. Aluminum is generally corrosion resistant, but it is damaged by most common cleaning agents *and by beer.* Therefore aluminum kegs are internally coated with paraffin. This requirement makes aluminum unsuitable for closed-system kegs, such as the Sankey, but it has been used in open-system kegs (Hoff-Stevens and Golden Gate), which do not require automated filling or cleaning.

c. Coated mild steel.   Coated mild steel is not commonly used as a keg material, but it is used in kegs designed for one-way, one-use export trade, in which return to the brewery is not cost-effective. The vessel is coated with a material that prevents any exchange between the product and the mild steel interior. The exterior is usually painted to protect against corrosion. These kegs are more dent-resistant than aluminum kegs. Almost all of these kegs are of the closed style, so that it is difficult to inspect them for damage to the coating if they are used repeatedly.

d. Stainless steel.   The most common and desirable material in modern keg production is stainless steel. It is suitable for both open and closed systems. Once it has been properly pacified ("pickled"), stainless steel is relatively inert in contact with beer. Stainless steel kegs can be cleaned with all cleaning chemicals commonly used in the brewery. It is often easy to repair these kegs. They are relatively light but very durable and strong. It is not unusual to find stainless steel kegs that have been in the trade for 50 years. Stainless steel kegs are often the most expensive, but the cost should be weighed against their long service life.

e. Variants.   One variant is the polyethylene or vulcanized rubber over metal keg. This design is typically used for closed-system kegs, but

Figure 3.2. Kegs with rubber chimes (left) and with steel chimes.

not exclusively. The internal metal component is usually stainless steel. The advantage of this design is its superior insulating qualities and easy handling and stacking. The polyethylene or vulcanized rubber protects the metal component from some damage, and metal of a lighter weight can be specified, to reduce the cost. The drawback is that the coatings are easily damaged, and dirt and bacteria can hide in damaged areas.

In another variant, only the chimes of the keg (the rims at the top and bottom) are coated with polyethylene or vulcanized rubber (*Figure 3.2*). The chief advantage is resiliency in the most commonly damaged component of the keg. The disadvantage is that the coating can easily be cut, and damaged areas create lodging points for soil. In extreme cases the material can become detached from the body of the keg.

A third design is polyethylene containers used as small take-home packages. Plastic packages have cleaning and structural problems, such as cracking and scratching, which limit their service life. However, their small size and convenience have made them popular packages, especially with brew pubs.

Figure 3.3. Golden Gate kegs, showing the bunghole (left) and the tapping valve (right).

6. What are Sankey, Golden Gate, and Hoff-Stevens kegs?

Golden Gate and Hoff-Stevens kegs are open-system kegs, while the Sankey is a closed-system keg. The names refer more to the tapping system than to the keg itself. These kegs can be constructed from most of the materials described above.

a. Golden Gate keg. The Golden Gate keg is now rare, although it was common on the West Coast 20 years ago. It has a characteristic barrel shape and a bunghole for filling (*Figure 3.3*). It has two valves—one on the side near the bottom of the keg, for dispensing the product, and one on the top of the keg, for injecting gas. There is a small sump leading to the lower valve. The two valves are typically identical in construction. The bottom valve can be fitted with a small snorkel intended to empty the keg more completely. The valves are held in place with a threaded brass ring. Some older valves are threaded directly into the keg. The threads can easily be damaged by impact to the keg, but it is possible to repair them.

The valve employs a rotating disk with a sealing washer. The disk is most commonly semicircular or shaped like a bow tie, but it is sometimes round, with a small port for tapping. The disk is engaged by the tapping apparatus and turned 90 degrees to open. The sealing washer and O-ring should be periodically replaced, although this regular maintenance is sometimes neglected. Removal of the valve often reveals otherwise un-reachable soils. The automatic cleaning equipment available for these kegs was not designed to clean through the valve, as should be done. Thorough cleaning can be accomplished manually by employing a tapping device attached to a detergent pump by a hose. Attaching the device and running detergent through the valve would contribute to superior cleaning. However, it is very dangerous to do so, as it places an operator very close to pressurized detergents and relies on the operator's careful attention to avoid mishaps (remembering to turn off the pump before disengaging, properly attaching the tapping device to avoid leaks, attending to the condition of hoses so they don't burst, etc.). The best and safest method of cleaning is removal of the valve, complete disassembly, and detergent soaking.

Some brewers prefer Golden Gate kegs for traditionally drawn cask-conditioned beers, because of the two-valve arrangement. The upper valve is opened to the atmosphere, and the beer flows out of the lower valve under the force of gravity.

A brewer who intends to distribute Golden Gate kegs outside the brewery or pub should first find out whether distributors and accounts can still work with these kegs.

b. Hoff-Stevens keg. The Hoff-Stevens keg is similar to the Golden Gate, with one major exception: a single valve is used for tapping. This keg was once popular on both coasts but is less common today. It is still used, although primarily by old-guard regional and craft brewers. An externally threaded fitting containing the valve is welded to the top of the keg. The valve has two ports, each blocked with a separate spring-loaded ball. The tap utilizes two short spears, which open the check valves as the tap is fixed to the keg. The larger port is for dispensing the product, and the smaller port is for gas. A plastic tube is connected to the product port and extends to the bottom of the keg. This tube must be flexible in order to avoid damaging contact with the filling device. Even with great care, the tube can be damaged during filling, so that tapping will be impossible.

Like the Golden Gate keg, the Hoff-Stevens keg was not designed for automatic valve cleaning. Both types of valves are susceptible to the same accumulation of contaminants unless they are serviced regularly. The plastic material used for the dispensing spear in the Hoff-Stevens keg does not hold up well to steam sterilization.

Some brewers have modified Hoff-Stevens kegs to convert them for use with a Sankey valve. This modification has advantages and disadvan-

Figure 3.4. Sankey keg.

tages. The valve design is more sanitary and sturdy, but the barrel-shaped keg is not completely compatible with the Sankey valve. The bung must still be removed for draining and cleaning, because the replacement valve does not occupy the lowest point of the keg when it is inverted. *It is impossible to completely remove cleaning agents through the Sankey valve alone.* The rigid valve spear will also interfere with the insertion of a conventional racking arm, unless the valve is relocated to an unconventional position.

c. Sankey keg. The Sankey keg (*Figure 3.4*) has become the keg of choice for most brewers. First introduced in the 1960s, it became widely used by the late 1970s. There are some variations in the valve in different parts of the world, but the general function is the same. In the United States, the Sankey system is the most common configuration, with the English Grundy system as an occasional variant.

Sankey kegs are identifiable by their straight sides (there are exceptions) and sturdy chimes. The upper chimes are formed with holes for easier handling. Most Sankey kegs are made of stainless steel. The valve is centrally located on the upper portion of the keg. It has a spear that reaches nearly to the bottom of the keg. The keg and valve are designed for fully automated cleaning and filling, with great improvements in package stability and beer quality. Sankey kegs are easily maintained and serviced, since they are widely available and are still being manufactured.

7. What sizes of kegs are used?

Common keg sizes are 15.5 gallons (1/2 barrel), 7.75 gallons (1/4 barrel), 5.16 gallon (1/6 barrel), and 5 gallons. These are not the only sizes available, but the use of other sizes leads to confusion if the product is sold outside the brewery. It is more work for tavern owners to determine the pour cost of a product in a nonstandard container and compare it to the cost of other products in kegs of standard sizes. Some brewers use 31-gallon (1-barrel) kegs, but they are too heavy for conventional distribution.

Some brewers use British keg sizes, such as the firkin and hogshead. These sizes are novel and create interest in the product, but it is difficult to distribute them in the trade.

8. How are open-system kegs cleaned?

a. Exterior washing.　The exterior of the keg should be cleaned before the bung is removed, to prevent soil from entering through the bunghole.

The bung must be removed in order to clean the interior of an open-system keg. Removal of the bung is most easily accomplished with a device designed for this purpose. A wooden bung can be removed by more primitive means (hammer and chisel), but there is a risk of damaging the sealing surface on the keg, and parts of the bung can be pushed into the keg, so that they must then be removed. Plastic bungs are more difficult, but effective tools for removing them are available.

b. De-ullage.　Old beer is drained from the keg. The draining of the keg is called de-ullage. The keg is then soaked or rinsed with clean water. Cold or warm water may be used, but it is inadvisable to rinse with hot water, which can cause some soil to bake on and become more difficult to remove. Soft water is preferable for all cleaning operations, in order to avoid the formation of mineral deposits. An advantage of automated systems is the ability to perform this step with rinse water returned from the washing step; not only is the water used twice, but it contains some diluted detergent.

c. Internal washing.　The keg is then ready for an internal detergent wash, commonly with a solution of either sodium hydroxide or phosphoric acid or both. The keg can be soaked or sprayed. Cleaning temperatures range from 120 to 150°F (50 to 65°C). Caustic and acid cleaners are usually mixed with other substances, such as wetting agents and surfactants, to help clean the keg and prevent redepositing soil on its surfaces. A cleaning with both caustic and acid is the most desirable. If this is not practical, regularly alternating cleaners over time will help prevent buildups of protein and beer stone. Concentrations of both caustic cleaners and acid cleaners range from 0.5 to 1.5% by weight.

Chlorinated compounds are highly effective against protein buildup. However, they are highly corrosive, even to stainless steel, in an acidic environment. Carbon dioxide, commonly used as a dispense gas and present in empty kegs, creates such an environment. Every effort should be made to evacuate the keg with fresh air prior to cleaning. *Never use a chlorinated cleaning product unless you are absolutely sure there is no carbon dioxide present.* Additionally, in the presence of water, carbon dioxide can neutralize sodium hydroxide, because it forms carbonic acid (H_2CO_3), and thus considerably diminishes the effectiveness of the cleaner. Acid cleaning alone is ineffective in removing many types of soil commonly found in kegs.

If the keg is constructed of a material other than stainless steel, cleaning compounds should be selected with caution. The best advice would be to consult your chemical supplier.

d. Rinsing. Detergents should be rinsed with clean water of a temperature similar to that of the detergent cycle. Then a final rinse with 180°F (80°C) water completes the washing cycle. Soft water is particularly beneficial for the final rinse. Periodic checks of the rinse water pH will ensure complete rinsing and chemical purity.

9. How is steam used to sanitize open-system kegs?

If an adequate supply of saturated steam is available, kegs should be steamed for approximately 60 seconds. However, this is almost impossible with a manual cleaning system and is also dangerous. Additionally, although this practice can ensure that a keg is sterile at the time when it is steamed, there is no assurance that an open-system keg remains sterile once it has been handled. It is advisable to visually inspect the interior of open-system kegs before filling, to check for residual soils and off-smells.

10. How are closed-system kegs cleaned?

Closed-system kegs are cleaned in a similar manner, but with some important differences. Although they can be cleaned manually or with semiautomatic equipment, many of the advantages of this type of keg will be lost unless an automated cleaning and filling system is used. Investment in a cleaner-filler specifically designed for this purpose is highly recommended. Modern cleaning equipment incorporates many safeguards necessary for maintaining the quality of the product in a vessel that is not visually inspected.

Closed-system kegs are turned upside down for internal washing and filling. The reason lies in the novel design of the dispense valve. During wash cycles, the product dispense port serves as the point of entry for water, detergent, steam, and gas, and the gas port serves as a drain. The importance of cleaning in this manner must be stressed. The cleaning solution is forced up the product tube and is sprayed onto the internal surface of the keg at high velocity. During this operation the cleaning solution should be pulsed from high flow to low flow, in order to properly clean the exterior surface of the valve spear. Without this pulsing, the spear will not be cleaned.

a. External washing. Closed-system kegs can be externally cleaned with little regard for introducing soil into the interior, because of their sealed valve configuration. The exterior of the keg can be cleaned by various means, ranging from manual labor to automated tunnel washers incorporating brushes and limited internal prewashing. The keg is then ready for internal washing.

b. De-ullage. Sterilized air is ported to the product valve of the upside-down keg. Any contents remaining in the keg are routed through the gas valve and dispensed to a receptacle or drain. Automated equipment senses the completion of this operation with a liquid sensor. It is not desirable to heat the keg at this stage. Heating can cause soils to become baked onto the surface, making subsequent cleaning more difficult. The keg is then rinsed with water or, preferably, with a diluted detergent solution captured from rinse water used in the washing cycle. Ideally, the rinse temperature should not exceed 120°F (50°C). The vessel should be gradually warmed over subsequent cycles to avoid unnecessary shock. Gradual warming also helps to avoid fixing soil to the surface.

c. Internal washing. Once rinse water is drained, the detergent cycle begins. The cleaning cycle for closed-system kegs is similar to that of open-system kegs, described previously, with the primary difference being the means of entry and exit from the keg. The flow rate is varied, as in the first rinse cycle. The detergent is evacuated from the keg with air or steam. *It is critical to employ a reliable system of detection for complete evacuation of cleaning solutions.* Cycle times vary amongst brewers, but

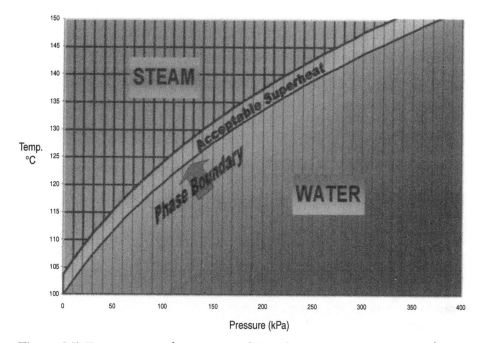

Figure 3.5. Temperature and pressure conditions for steam.

with a properly working automated system, the cycle time can range from 45 to 60 seconds. Single- and dual-detergent systems are available, with dual systems being more desirable.

  d. Sterilization.   The cleaned keg is then ready for sterilization. This is perhaps the most important advantage of the closed system over the open system. Closed-system kegs can be safely steamed to a temperature well above the survival threshold of common beer spoilers. Automated cleaning systems require a source of clean saturated steam, free of boiler treatment chemicals and minerals. Steam is saturated under pressure and temperature conditions at the phase boundary at which liquid water turns into gaseous steam (*Figure 3.5*). Saturated steam, or wet steam, must be held under controlled pressure and temperature for sterilization. Pressurized steam has about five times the energy of water at the same temperature. As steam condenses on the keg walls, this additional energy is delivered to the keg surfaces. Condensing, wet steam delivers the heat energy necessary to destroy beer spoilage organisms quickly (*Table 3.1*). Wet steam also helps prevent the formation of mineral scale on surfaces inside the keg, if the rinse water is less than perfectly soft. The steam also evacuates much of the air from the keg prior to filling, thus

Table 3.1. Sterilization times for destruction of beer spoilage organisms

Saturated steam		Dry, superheated steam	
Temperature	**Time**	**Temperature**	**Time**
100°C (212°F)	20 hr	120°C (248°F)	8 hr
115°C (239°F)	2.5 hr	140°C (284°F)	2.5 hr
120°C (248°F)	50 min	160°C (320°F)	1 hr
125°C (257°F)	6.5 min	170°C (338°F)	40 min
130°C (266°F)	2.5 min	180°C (356°F)	20 min
135°C (275°F)	1 min		

providing another benefit. None of these benefits can be obtained with open-system kegs.

The keg should then sit for one complete machine cycle (at least 60 seconds) to receive sufficient heat treatment. A good rule of thumb is to select a temperature comfortably higher than that at which the most stubborn beer spoilage organisms can survive, 275°F (135°C), and hold that temperature throughout the keg for 60 seconds.

e. Evacuation.   Next the keg valve can be briefly rinsed with steam to ensure a sterile connection to the keg if it has been moved to another station. The keg is then flushed with an inert gas, such as carbon dioxide, to remove any condensate. Precautions must be taken to ensure that the counterpressure gas is sterile. Many sterilizable filters are available for this purpose.

f. Counterpressure.   The keg is then prepressurized with gas prior to filling. Air is not recommended as a flush or counterpressure gas. The amount of dissolved oxygen in the packaged product can be kept very low if an inert gas is used. This is another advantage of the closed-system keg, as it is nearly impossible to match this low level of dissolved oxygen in an open-system keg.

11. How are open-system kegs filled?

Open-system kegs have the advantage of inherent simplicity. They can be filled by simply inserting a piece of hose from the finished beer tank into the bunghole and opening a valve. This method is recommended only for low-carbonation, cask-conditioned products. Filling by this method results in a high level of dissolved oxygen in the beer, often many times greater than that of the most poorly packaged bottled beer. *It is a myth that because a keg can be filled to the brim there will not be any oxygen in the package.*

Figure 3.6. Racking arms.

12. How does a racking arm operate?

The proper method of filling an open-system keg is to use a racking arm (*Figure 3.6*). A racking arm allows the keg to be sealed against a rubber stopper. Counterpressure gas can be employed, so that the keg can be filled with a fully carbonated product. Beer enters the pressurized container either by pump or, from an overhead tank, by gravity. There is no pre-evacuation of air from the keg. Using an inert counterpressure gas is good practice, but it does not significantly lessen the ingress of oxygen during filling. The product enters the keg in a somewhat more controlled manner than by the primitive method of filling from a hose, and therefore the keg filled by a racking arm may contain less dissolved oxygen. Neither method allows filling the keg to anything less than nearly brim-full. (It should be noted that the actual volume of a keg is greater than its stated volume, to allow for minor dents incurred in the trade. A well-used keg should still maintain its full stated volume.) Incorporating a scale beneath the filling keg allows the operator to fill the keg by weight rather than volume. If you are giving away an extra 1/4 gallon of beer with each half-barrel, you will lose a full half-barrel after filling only 62 kegs. This method of filling burdens the operator and may limit the number of kegs that can be filled at the same time by a single operator.

With the general acceptance of the closed-system keg, it has become more difficult to find and maintain racking arms.

13. How are closed-system kegs filled?

Closed-system kegs can be filled manually, by means of a converted tapping device or a proper filling valve, but primitive methods (such as removing the valve, filling the keg, and reinstalling the valve) are not recommended. Without the use of flowmeters or checkweighers, closed-system kegs can be overfilled, just as open-system kegs can.

In an automated system, closed-system kegs are inverted for filling, as in cleaning. The keg is clamped to the filling station, and the valve is rinsed with hot water or steam. The valve is then flushed with inert gas, to remove any excess water or air. The keg is again flushed briefly with inert gas, to remove any condensate. It is then counterpressured to within a few pounds per square inch of the product pressure at the point of delivery. The best filling performance can be obtained with the use of a product pump. Product pressure must always be maintained at or above the gas saturation point of the product. At this point the dispense gas valve of the keg and the product valve of the racker can open. The keg valve has a separate rubber-seated port to allow dispense gas to enter the keg and a ball valve to allow the product to exit when the keg is tapped. During the filling process the functions of these two valves are reversed. This unique valve design allows the beer to enter the lowermost portion of the vessel first, in a calm, unagitated fashion. Counterpressure gas is evacuated in a metered fashion through the spear and ball valve to the atmosphere. Closed-system kegs should not be filled through the normal product port.

A filling system in good working order should fill a 1/2-barrel keg in 60 seconds or less with no gain or loss of carbonation. The dissolved oxygen level should be less than 20 parts per billion. With this method of filling and the utilization of flowmeters or checkweighers (or both), filling efficiency of 98% or more can easily be achieved with little or no operator intervention.

14. How is the draft beer package sealed?

Open-system kegs and casks are sealed with a bung, driven into the opening on the side of the keg (*Figure* 3.7). The most common type is made of compressed poplar wood. Plastic bungs are also suitable.

Figure 3.7. Driving home the bung in a Golden Gate keg.

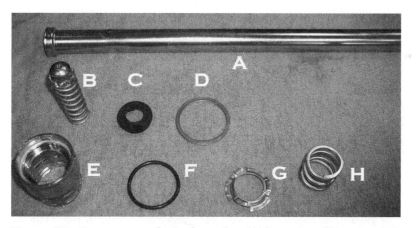

Figure 3.8. Components of a Sankey valve: (A) keg spear, (B) spring with poppet, (C) seal, (D) circlip, (E) valve body, (F) O-ring, (G) retaining ring, and (H) spring.

Closed-system kegs are sealed automatically by a spring-loaded ball valve and a rubber-seated gas valve incorporated in the dispense port. The components of a Sankey valve are shown in *Figure 3.8.* The valve seats require occasional maintenance. Torn or damaged valve seats will cause

Figure 3.9. Sankey valve repair tool.

excessive foam during tapping. Rebuilding a valve is simple with the proper tools (*Figure 3.9*).

15. Why should the interior of open-system kegs be inspected?

Open-system kegs should be inspected to ensure that they have been properly cleaned, that bung remnants have been removed, and that residual liquid has been completely drained away. Specialized light sources for keg inspection are available.

16. How are closed-system kegs inspected?

Closed-system kegs do not require inspection with each fill, but routine periodic inspection is recommended, to check the effectiveness of cleaning operations. A good inspection procedure is to randomly select kegs after cleaning, remove their valves, and audit them for cleanliness.

17. What is cask conditioning?

Cask conditioning is the process of maturation of beer or ale in the keg or cask. This was once the only method of maturation and carbonation of packaged beer. Cask conditioning has almost disappeared in Great Britain, but the practice has been revived in England and in the United States, primarily by specialty and craft brewers of ales and stouts.

a. The cask.   The first consideration in cask conditioning is the selection of a suitable vessel. Depending on the desired final carbonation level, any suitable material can be used, including wood. Classic cask-conditioned beers typically vary in carbonation from 1.8 to 2.2 volumes, but there are exceptions on either side of this range. Traditional English-style beers run toward the lower end of the range. Exotic containers, such as wood and aluminum casks, get more attention at the bar but require more care and preparation. Most open-system kegs work well for the addition of dry hops, finings, or primings. Traditional cask designs feature a hole (the keystone) at the cask head and a bunghole on the side, for both cleaning and filling. After a cask has been cleaned, a small wooden bung is pounded into the keystone.

b. Racking.   Traditionally, finished beer is racked directly from the fermentation tank to the cask, without filtration. However, many brewers pass the beer through at least a rough filtration to improve clarity. Beers with low levels of carbonation are racked directly into the cask without counterpressure and bunged. In the center of the cask bung is a plastic tut, which will later be removed for tapping.

c. Additions.   Fresh hops or finings are commonly added at racking, as is priming sugar. Finings are traditionally composed of isinglass (fish bladder protein) or gelatin, together with auxiliary finings that aid these materials. Finings are designed for yeast removal and do almost nothing for chill stability. However, cask ale is normally served relatively warm (50–55°F, or 10–13°C) and thus is not subject to the formation of chill haze. Fresh whole hop cones or slurried pellet hops add a distinctive dry-hop flavor. Priming sugar solutions are typically boiled first and added hot, either directly to the cask or to the tank from which the beer is racked. Cane sugar, dextrose, fresh wort, or syrups can be used as priming sugars, depending on the brewer's preference.

d. Conditioning.   The vessel must be capable of handling the additional pressure caused by secondary fermentation over a reasonable temperature range. Bourbon and wine barrels are not designed to withstand the pressure buildup and should not be used without a ventilation device, such as a fermentation lock. Viable yeast must be present in the product. All operations, including maturation, should be carried out at a temperature at which refermentation can proceed, 55–65°F (13–18°C).

Ale takes at least seven days in a primed and sealed cask in the cellar to become fully conditioned.

e. Stillage and tapping.   The cask should be set in a special cradle or stillage rack in which it is slight propped forward, with the bung facing up and the keystone at the front. When the cask is settled and ready, a hard peg, or spile, is pounded through the plastic tut, in the center of the bung, creating a vent hole. The cask may then be allowed to breathe out excess carbonation. Once the cask is deemed ready for tapping, a brass or plastic tapping cock is pounded through the wooden keystone, allowing beer to be emptied from the cask either by gravity dispense or beer engine (see Chapter 3, Volume 3). Some trial and error is necessary to determine the exact amount of priming sugars necessary for each beer. A well-conditioned ale will release carbonation at tapping but not gush. A properly conditioned beer will exhibit a gentle effervescence when swirled in the glass, without being overly "gassy."

Closed-system (Sankey) kegs can be used for cask conditioning, with the precaution that a more thorough inspection of the keg, involving removal of the valve, may be required after emptying. When fresh hops or finings are added to the keg, there is a risk of clogging the valve or tap.

In pub operations, these techniques can be carried out on a larger scale, using small serving tanks with appropriate pressure-handling capability to supply the bar directly.

18. What methods are used to identify kegs in the trade?

Kegs that are to be circulated through a brewery's normal distribution channels should be uniquely identified so that they can be returned to the brewery. Kegs can be identified with metal nameplates, embossed chimes, vinyl stickers, and painted striping. Colored stripes painted or taped around the keg help to quickly identify the brewer. However, it is common for unsuspecting brewers using similar color schemes to receive misdirected kegs. There are no standards for keg identification, but a national registry of keg identification is maintained by the Brewers Association of America (BAA). Bar codes can also be used to help identify kegs.

Keg identification should not be confused with product identification. Uniquely identifying kegs by product can lead to an unruly float unless it is necessary (for example, to distinguish kegs used for soft drinks and kegs used for fruit-flavored beers in the same operation).

Figure 3.10. Ink-jet printer applying product identification code on a keg.

19. What methods are used to identify the contents of kegs?

The contents of kegs can be identified in numerous ways, provided that the labeling meets legal requirements. Self-adhesive labels, printed rings attached to the neck, and ink-jet printing directly on the keg (*Figure 3.10*) are all common. Plastic cap seals for valves can also be used to identify the contents. Wooden bungs may be stamped or branded to identify the contents (*Figure 3.11*).

Anything that can be removed easily from the keg will not make a satisfactory nameplate.

20. How do you fill-check a keg?

The two most common methods of fill-checking are (a) passing the product through a flowmeter while filling the keg and (b) physically weighing the full keg. Most operations use both methods.

Checkweighing becomes more difficult when there is a diverse float of kegs from many manufacturers or kegs made of dissimilar materials present in the brewery. Fill level inspection is difficult for similar reasons.

21. What constitutes a sound sanitation program?

The same good manufacturing practices employed in the rest of the brewery are also applicable on the racking line. The ability to supply hot

Figure 3.11. Stamping bungs with an identification code.

water (180°F, or 80°C) is essential for hot-sanitizing and rinsing kegs. Supplies of cleaning agents will be required, in the proper concentration and at the proper temperature. Pumps will be necessary to circulate cleaners and sanitizers through the operation. The ability to periodically disassemble and inspect product piping, filling heads, and cleaning heads is important. Good maintenance and documentation will yield excellent results. Clean-in-place regularly. Hot sanitation before each use is highly recommended.

22. What are the advantages of automation?

Automation yields at least two major benefits. First, it reduces labor, both in terms of the number of people required and the physical difficulty of doing the job. Second, it improves quality. Automated tasks can be performed repeatedly with little or no variation. Quality can be monitored, and records can be generated easily.

Automation should be regarded as an enhancement of the human element, not as a replacement for it. Freeing an operator from the physicality of the job should allow the operator more time for quality checks.

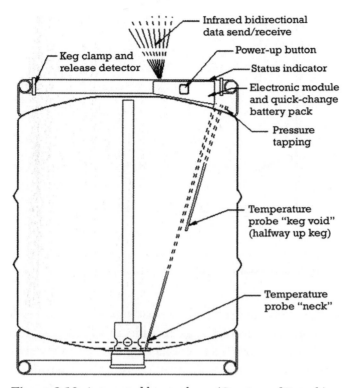

Figure 3.12. Automated keg analyzer. (Courtesy of Rotech)

23. What is a good quality assurance routine for a draft operation?

Regular checks should be made for microbiological cleanliness of the product, storage tanks, piping, and filling equipment. Individual kegs should also be tested, in order to ensure adequate cleaning and sterilization.

Product should be sampled and tested for compliance to specification for alcohol content, carbonation, color, and gravity. Practical experience has shown the value of not only checking at typical sample points, but the finished package as well. Sampling devices can easily be fabricated from tapping equipment. Destructive package testing should be considered a necessary cost of doing quality business.

A brewer using closed-system kegs should have a test keg, with one or more sight glasses or portals, a thermometer, and a pressure gauge. Test kegs can be purchased, or a primitive test keg can be fabricated from an ordinary keg. One of the most useful innovations to come along in the last 20 years is the electronic test keg (*Figure 3.12*), which records tem-

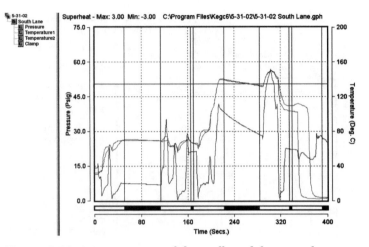

Figure 3.13. Screen capture of data collected from an electronic test keg. (Courtesy of Rotech)

perature and pressure and sends these readings to a computer for analysis (*Figure 3.13*). Electronic test kegs provide excellent monitoring of the cleaning and filling performance of a racking line, and they also can aid in diagnosing problems and fine-tuning the system for maximum performance and speed. Temperature and pressure readings can be used to determine if the steam supplied to kegs is saturated, an enormous benefit for disinfection. Most of the computer software for these kegs displays saturated steam conditions and accounts for the time spent at critical temperatures.

REFERENCES

U.S. Code of Federal Regulations, Title 27 (Alcohol, Tobacco Products and Firearms), Chapter I (Alcohol and Tobacco Tax and Trade Bureau, Department of the Treasury), Part 7: Labeling and Advertising of Malt Beverages.

U.S. Department of the Treasury, Alcohol and Tobacco Tax and Trade Bureau. 2001. *The Beverage Alcohol Manual.* Volume 3: *Basic Mandatory Labeling Information for Malt Beverages.*

Vogel, Edward H., Jr., Schwaiger, Frank H., Leonhardt, Henry G., and Merten, J. Adolf. 1946. *The Practical Brewer: A Manual for the Brewing Industry.* Master Brewers' Association of America, St. Louis, Mo.

CHAPTER 4

Bottle Shop Operations

Stephen Bates

BridgePort Brewing Company

1. What is the general bottle shop procedure?

Bottle shop procedure consists of the following operations:

a. Receiving new bottles, either in bulk or in shipping containers from the glass factory
b. Receiving cans in bulk from the aluminum can factory
c. Rinsing new bottles or cans
d. Filling and crowning bottles and filling and seaming cans
e. Pasteurizing bottled or canned beer (optional)
f. Labeling the bottles
g. Erecting knockdown cartons for packing bulk glass bottles
h. Packing bottles and cans into cartons and sealing the cases
i. Code-dating cases for quality assurance identification
j. Palletizing cases for shipment

2. What is a finishing cellar?

The finishing cellar is the room in which the final filtered beer is stored prior to packaging. Beer is pumped from the cellar to the filler.

3. What types of beer pumps are used?

There are generally three types of pumps used for transferring beer from the finishing cellar to the filler:

a. Positive displacement (PD) pumps provide constant pressure with little negative effect on the beer. They require a pressure relief or

Figure 4.1. Rotary rinser cleaning bottles prior to filling. (Courtesy of BridgePort Brewing Company)

bypass system to relieve pressure when the pump is running and no beer is being called for. Most fillers are made with level sensors to turn the pump on or off.

b. Centrifugal pumps provide constant pressure while running but may mechanically shear the beer when they are pumping against a closed inlet valve or high pressure.

c. Diaphragm pumps work off compressed air. They work well on slow-filling lines and gently pump at high counterpressure, but they tend to freeze up when operated at high speeds.

4. How are bottles sterilized for filling?

New glass bottles and cans are relatively clean when they arrive from the factory and need only to be rinsed to remove dust and debris (*Figure 4.1*). A food-grade sanitizer added to the rinse water may be helpful. Glass that has been wet during storage should not be used, because it may have mold growth.

5. What are the primary systems used in the filling process?

Bottles are filled under carbon dioxide counterpressure. The beer flows by gravity into the container, which vents back into the filler bowl.

Figure 4.2. Short-tube bottle filler with crowner. (Courtesy of BridgePort Brewing Company)

Two systems are generally employed:

a. Long-tube fillers have a fill tube that reaches to the bottom of the container, which is filled completely from the bottom, with the tube providing the displacement necessary for headspace in the container. This method of filling is very gentle and reduces air pickup, even without pre-evacuation of the container.

b. Short-tube fillers (*Figure 4.2*) are designed in such a way that, when the fill valve is opened, beer floods evenly down the bottle sidewall. As the bottle is filling, gas inside the bottle is displaced through the vent tube and into the filler bowl. As the bottle continues to fill, the product reaches the vent tube, and then travels up the tube. When the beer in the tube reaches its level equilibrium with the filler bowl, the filling stops.

6. What are the important factors in filling a container with beer?

Eight important factors affect filling a container with beer:

a. Counterpressure.   Carbon dioxide counterpressure is used to maintain carbonation during the filling operation. Air counterpressure is not recommended.

b. Sanitation.   The cleanliness of all equipment and beer piping is especially critical in filling operations in which the beer is not pasteurized after filling.

c. Bottle or can fill level.   State and federal laws require fill volumes with close tolerances, and breweries are audited to make sure that they comply.

d. Air elimination.   Air must be removed from the headspace of the container, either by pre-evacuation or by quiet filling, to preserve the flavor and shelf life of the beer.

e. Quiet fills.   Filling should proceed with minimum turbulence and foaming, with either a long- or a short-tube filler.

f. Overflow capacity.   There must be enough room in the bottle to allow for the expansion of beer during pasteurization.

g. Beer temperature.   Temperature is critical. If the beer is too cold, it may not produce enough foam during the jetting process. If it is too warm, the beer may be too wild to control before crowning.

h. Quality checks.   At regular intervals, the carbon dioxide content, headspace air, and fill level in crowned bottles must be checked.

7. Why is carbon dioxide preferred for counterpressure in bottling tanks?

Carbon dioxide maintains gas in solution without introducing air into the beer as dissolved oxygen. Oxygen quickly reacts in the beer and causes the development of staling flavors.

8. What is the proper method for cleaning a beer filler?

The cleaning method depends on the type of filler. Some fillers can be cleaned by back-flushing beer lines, filler bowl, valves, and fill tubes in a recirculating clean-in-place (CIP) system. Some fillers require opening the filler bowl and cleaning by hand. In most cases, an alkaline cleaner in warm water is used, sometimes followed by an acid neutralizer, and then a sanitation step. The filler and lines should be cleaned when an operation is finished and within 24 hours of the next use. *Figure 4.3* shows a short-tube filler with CIP cups used in cleaning and sanitizing cycles.

Fillers should be sanitized just prior to use. Sanitizing can be achieved most effectively with hot (180–200°F, or 80–95°C) water circulation or with chemical sanitizers. Some fillers, designed to accommodate higher temperatures, can be sanitized with a steam-in-place (SIP) system.

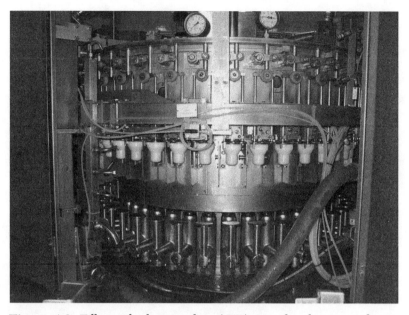

Figure 4.3. Filler with clean-in-place (CIP) cups for cleaning and sanitizing. (Courtesy of BridgePort Brewing Company)

9. Why is air detrimental to bottled beer?

Excessive air pickup during packaging has been shown to increase oxidation haze and stale flavors. The chemical reaction of oxidation is slow at ambient temperatures but much more rapid at pasteurizing temperatures. When bottled beers are stored, oxygen in the neck of the bottle is gradually absorbed by the beer, causing deterioration of the flavor. Air pickup during packaging is about as detrimental as microbe contamination and dramatically reduces shelf life

10. What precautions are taken to help eliminate air in the filling process?

Some fillers (single-evacuation and double-evacuation fillers, respectively) incorporate a one- or two-step pre-evacuation process, to eliminate air from the bottles prior to filling. In a single-evacuation system, the bottle is subjected to a vacuum and then pressurized with carbon dioxide. In a double-evacuation system, the process is repeated, so that there is effectively no air remaining in the bottle. The air is replaced by carbon dioxide, which helps to prevent the mixing of beer and air during filling.

To remove air from the neck of a bottle after filling, a device called a jetter injects a high-pressure pinpoint stream of water into the neck, strik-

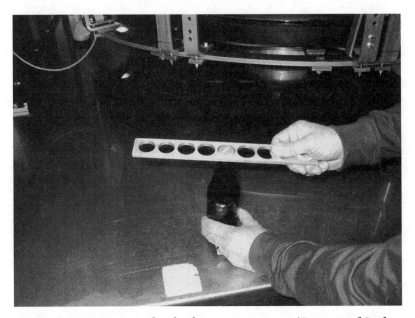

Figure 4.4. Crimp gauge for checking crown crimps. (Courtesy of Bridge-Port Brewing Company)

ing the beer and causing it to foam over, thereby driving the air out of the neck. The bottle is then immediately crowned. Headspace air, a combination of oxygen and nitrogen gases, is expressed as milliliters of air, according to procedures by Zaam and Nagel. Most modern fillers can deliver 12-oz. bottles with less than 0.20 ml of headspace air.

11. How are bottles sealed?

There are generally three types of closures: pry-off and twist-off crowns and roll-on caps. Pry-off and twist-off crowns are applied by a machine called a crowner, in which rotating heads move down over the crown and crimp it onto the bottle. Roll-on caps are applied by a machine called a capper, in which rotating disks conform the cap to the screw threads of the bottle and rollers tuck the cap pilfer ring to the bottle.

12. How are bottles checked for crown crimp?

After a bottle has been crowned, the crimp size is checked with a crimp gauge, a narrow metal bar with predrilled holes, the sizes of which are marked on the gauge (*Figure 4.4*). The hole that best fits over the crown is selected. Generally, the crimp size should be between 1.025 and 1.135 in., depending on the finish of the bottle and the thickness of the crown shell.

13. What is the crown shell made of?

The crown shell is made from tinplate or tin-free steel plate with a protective finish.

14. What are crown liners made of?

Crown liners are made of polyvinyl chloride formulated to provide low opening torque. Some liners have oxygen-scavenging properties.

15. What precautions are necessary in crowning?

The following precautions are basic:

a. The crowner bottle star wheel must be properly aligned with the crowning heads.
b. The crown throats must be in good condition and all worn throats replaced.
c. Proper crowning pressure must be applied.
d. Crowns must be crimped to proper specification and checked with a crimp gauge.
e. Twist-off crowns must be checked with a torque gauge.
f. The crowner must be thoroughly cleaned to prevent contamination.

16. What is pasteurization?

Pasteurization is the partial sterilization of certain fluids by heating to temperatures of 130–160°F (55–70°C), named for the French scientist Louis Pasteur, who demonstrated this effect in 1864. Beer is a poor medium for the growth of most microorganisms, because of its low pH and the presence of alcohol and hop flavoring, which inhibit growth. However, some nonpathogenic microorganisms can survive in it. Beer is pasteurized to increase its shelf life by killing spoilage bacteria, wild yeast, and molds.

17. What additional effects does pasteurization have?

The heat necessary for pasteurization causes the oxidation of various organic compounds, which is detrimental to the flavor of the beer. Heat may also cause coagulation of nitrogenous substances, which leads to the formation of haze. Some caramelization can also occur with excessive heat or prolonged holding time, giving the beer a cooked flavor.

18. What mechanical means are used to pasteurize beer?

Bottled or canned beer is passed through a machine called a pasteurizer, which contains tanks of water heated by either heat exchangers or steam coils. As the containers enter the pasteurizer, the beer is brought up to critical temperature for pasteurization. The containers are then allowed to gradually cool until the beer reaches room temperature.

19. Why is rapid heating and cooling detrimental to successful operations?

Very rapid heating and cooling can cause a high percentage of breakage of bottles and can result in poor efficiency. Very slow heating and cooling may increase the effects of oxidation and affect the flavor and brilliance of the beer. Heating and cooling canned beer is less troublesome, because breakage does not occur, and because metal cans conduct heat more effectively than glass.

20. What is the critical temperature for pasteurization, and how long must beer be held at this temperature?

The generally accepted practice is to pass bottles of beer through a water bath or under sprays at the "critical" pasteurizing temperature for about 10 minutes. Treatment for 8 minutes at 140°F (60°C) usually ensures satisfactory pasteurization. Pasteurization conditions are measured in pasteurizing units (PU):

$$1 \text{ PU} = 1 \text{ min at } 140°F$$

To achieve the required residence time, pasteurizers require large amounts of conveyor space and operate at slow speeds.

21. What happens if the pasteurization temperature and time are not properly controlled?

Insufficient or excessive pasteurization produces unstable beer with poor appearance and flavor.

22. What means are used to check the pasteurization temperature and time?

A device called a traveling recorder is sent through the pasteurizer to record the time and the temperature of the container. The results are expressed in pasteurizing units.

23. What types of pasteurizers are available to the brewing industry?

Generally, two types of pasteurizers are used:

a. Tunnel-type pasteurizers are used after filling and before labeling. They have heat exchangers or steam coils to heat water, which is sprayed over bottles on a conveyor. The pasteurizer is made up of several heating sections. Bottles enter a section in which they are slowly heated, and they pass from one section to another, each at a higher temperature, until the beer reaches 140°F (60°C). It is held at this temperature for approximately 5 to 8 minutes, and then the bottles pass through sections at decreasing temperatures until the beer is cooled to approximately 70°F (20°C).

b. Flash-type pasteurizers are used before the beer enters the filler. They heat the beer rapidly, with heat exchangers and heating media (usually steam) creating higher temperatures and requiring a shorter residence time than tunnel-type pasteurizers. After the beer passes through the flash pasteurizer, it goes to a tank and then to the filler. The flash system requires contamination-free filling. Because of the rapid temperature changes, off-flavors are less likely to develop than in beer pasteurized in a tunnel system.

24. Can beer be sterilized by means other than a pasteurizer?

Beer can be microbiologically stabilized by sterile filtration, in which it is passed through membrane filters with ultrafine pores (0.45 microns). Sterile filtration systems commonly use depth filters, canister-style filters, and plate filters.

Filtration is performed under tightly filtered (MERV-14) filtered air that is supplied over the filler. The filtered air creates a positive pressure in an enclosed filling operation and helps prevent dirty, unfiltered air from reaching the product containers.

25. How are the labels applied?

Labeling machines are usually either glue-applied or pressure-sensitive roll-on types.

a. Glue-applied labelers, either rotary or straight-line types, have a label magazine, an adhesive applicator, and a brush or roll-down section. In the beer industry, glue-applied labels are more popular than pressure-sensitive labels, because of their relatively low cost and the high efficiency

Figure 4.5. Labeler with front and rear aggregates. (Courtesy of Bridge-Port Brewing Company)

of glue-applied label machines. A glue-applied labeler capable of applying front, neck, and back labels is shown in *Figure 4.5.*

b. Pressure-sensitive labelers use labels with a preapplied adhesive. The labels are peeled from the backing liner and transferred to the bottles by air jets and then smoothed with a wipe or roll-down to ensure good application. Pressure-sensitive labels are more expensive than glue-applied labels. Their advantages are excellent label print quality, ease of application, shorter setup time , and quick machine cleanup.

26. What are the key parts of a glue-applied labeler?

Labels are most commonly applied to bottles with glue by a series of steps involving the following equipment:

a. Glue pumps supply the glue to the glue roller. The adhesive may be heated (approximately 75–80 degrees), decreasing its viscosity and increasing its runability.

b. Glue roller, commonly made of rubber, holds the glue. A glue knife spreads the glue in an adjustable, even layer onto the roller.

c. Label pallet is a contoured piece of alloy machined to the precise shape of the label. The pallet rolls against the glue roller, picking up glue. It then rotates to the label magazine, where a label is transferred to it.

d. Label magazine holds the labels. The labeler operator loads the labels onto the magazine. Labels are picked up by the rotating pallet and transferred to the gripper assembly.

e. Gripper cylinders use fingers with a sponge backing to take a glued label from the pallet and apply it to a bottle.

f. Brushes and rollers smooth the label onto the bottle as it proceeds through the machine.

g. Code-dating jets or stamps are typically applied at the end of the labeling operation.

27. What types of adhesives are available for glue-applied labelers?

Generally, two types of adhesives are used. Both have excellent cold-water immersion capability.

a. Casein adhesives are based on milk protein. They adhere well to cold or wet bottles and clean up relatively easily with tepid water.

b. Noncasein (synthetic) adhesives usually do not have the grab quality of casein adhesives on cold or wet bottles. They are typically used in applications where bottles come out of the pasteurizer warm and dry. Generally, noncasein glues cost less than casein glues.

28. What precautions are necessary in labeling?

Labels can be placed on the face, neck, and back panel of a bottle. If neck and back labels are applied, they should be in proper registration with the face label.

The proper amount of adhesive must be used on glue-applied labels. To determine whether the proper amount has been applied, the label on a freshly labeled bottle can be peeled back to expose the glue pattern. The label should not have so much glue that it swims or can easily slide around on the bottle, and it should not have so little glue that it leaves a poor pattern on the glass. The label adhesive must be compatible with the surface treatment of the bottle (applied during the annealing process). A good-quality label adhesive machines well, cleans up easily, and adheres well to cold or wet bottles. It must also have excellent cold-water immersion capability, to withstand conditions in ice chests and store coolers.

29. What kinds of label papers are available?

All labels should be printed on paper with good wet strength, to help prevent them from breaking down when saturated with water. Commonly used label papers include

    a. Coated one-side label paper
    b. Metallized paper
    c. Foil-laminated paper
    d. Poly-laminated paper

30. What kinds of labeling problems should you look for?

Labeling machines should receive good preventive maintenance, and they must be timed and tuned occasionally so that the transfer points work in optimal sequence.

Label and bottle batch number data should be checked. Adhesive, bottle surface treatment, label grain direction (labels should be printed "grain long," that is, with the grain of the paper perpendicular to the height of the bottle), flagging, curling, wrinkling, and delaminating of metallized paper are a few things to look for.

31. How should labels be stored?

Bundled or shrink-wrapped paper and foil labels should be stored at 70–75°F (21–24°C) and 50–60% relative humidity. Unwrapped bundles of paper labels should be stored at 75–80°F (24–27°C) and 70–80% relative humidity.

32. What are label collation marks?

The collation mark is an inked line printed on one side of a bundle of labels to help the labeler operator identify the bundle direction. It ensures that the labels are not placed in the label magazine upside down or backwards.

33. Why do shape and colors affect the price of labels?

The shape and color of labels determine the number that can be printed on a sheet. Labels are usually printed and then cut out of a stack of 1,000 sheets at a time. Labels that are square or rectangular and have the same color on the edges can be cut once without wasting material. If the labels are irregularly shaped or have different border colors, fewer labels can be placed on a sheet, and the sheets have to be precut into squares or rectangles in stacks of 1,000 and then die-cut to their final shape. The number of colors also affects the price, since the cost of ink the cost of labor for color registration and finishing time are included.

34. What is a label protection panel?

Most bottles are made with a recessed body panel where the label is applied, located between the bottle contact areas. The panel helps prevent labels from rubbing against each other and causing damage as the bottles travel on conveyors.

35. How are cold, wet bottles labeled?

During pasteurization, the bottled beer warms and dries the bottle, so that it is relatively easy to apply glue and labels. Unpasteurized bottles, however, are cold and wet and present a challenge to the labeler machine. Several strategies are commonly employed in labeling cold, wet bottles:

a. **Water shower** removes beer residue from bottles after they have been crowned.

b. **Air knives** remove water from the surface of bottles just before they enter the labeler.

c. **Casein-based glues** are typically used, since they generally work effectively on cold, wet bottles.

d. **Brush-down time** is maximized.

36. What are the main materials used in making glass?

Sand, soda ash, limestone, and cullet (recycled glass) are used to make new glass bottles. The ingredients are mixed in furnaces at temperatures up to 2,700°F (1,500°C) until molten. As much as 70% cullet can be used in the manufacture of new glass bottles, and energy costs for glassmaking drop 0.5% for each 1% of cullet content.

37. Where does glass get its color?

Iron, sulfur, and carbon are added to make amber-colored glass. Chrome oxide is added to make green glass. Cobalt oxide is added to make blue glass.

38. What glass colors are used in beer bottles?

Generally, beer bottles are made in three colors: amber, green, and flint (clear).

39. What glass color is usually preferred?

Amber is considered to be the best color for beer bottles, because it minimizes the penetration of ultraviolet light, which has a detrimental effect on the flavor of beer.

40. Why do glass manufacturers apply a surface treatment to beer bottles?

Surface treatments are sprayed onto the glass just after the annealing (heat-strengthening) process, to form the surface to which the label glue adheres. Surface treatments provide several benefits:

a. They create a scratch-resistant and lubricious surface.
b. They help bottles to run more easily and handle better without binding while being conveyed.
c. They prevent abrasions on bottles, which could cause the bottles to explode, and thus ultimately surface treatments protect the consumer.

41. What types of surface coatings are used?

a. Nonpermanent coatings are organic materials, including stearate, polyvinyl alcohol, and sodium oleate. They can be washed off with water. These coatings are used on bottles that are not washed prior to filling.

b. Semipermanent coatings consist of a polyethylene treatment without an undercoating. These coatings make a water-resistant surface.

c. Permanent surface coatings consist of a polyethylene treatment applied over a tin oxide undercoat. These coatings do not wash off during pasteurization.

42. What processes are used in glass molding?

Glass bottles are made in steel molds, one forming the body and the other forming the top, or finish, of the bottle. Two processes are used to form the glass in the molds:

a. Blow-and-blow method is used for heavy bottles and return-for-rewash bottles.

b. Narrow-neck press-and-blow method is used for small-diameter finishes. This process is faster and can produce bottles of lighter weight, because the glass can be distributed more evenly.

43. What are the main parts of a beer bottle?

The parts of a typical stock beer bottle are shown in *Figure 4.6*.

44. What types of bottle finishes are available?

The finish, or the top of the bottle, holds the crown or roll-on cap. Each has a particular shape to hold the closure.

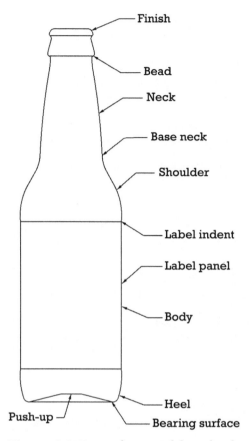

Figure 4.6. Parts of a typical beer bottle
(Courtesy of Owens-Illinois)

a. **Pry-off** finish crowns must be removed with an opener.
b. **Twist-off** crowns may be removed by hand.
c. **Roll-on** aluminum caps with pilfer bands are commonly used on 32- and 40-oz bottles and PET (polyethylene terephthalate) bottles.

45. What types of glass defects should you look for?

Typical defects in glass bottles are described in *Table 4.1.*

46. Why does each bottle have an identification code?

Each container can be identified by the cavity in which it was made. When automatic inspection equipment detects a defect in a bottle, it reads the identification code to determine which machine produced the

Table 4.1. Typical defects in glass bottles

Defect	Description	Affected areas	Possible consequences
Split finish	Vertical crack in top finish	Finish and seams	Leakage
Checks	Cracks	Finish, neck, base	Excessive breakage
Load lines, wash-boards, tears	Lines on surface of glass	Neck and body	Cosmetic damage
Blisters	Bubbles, air pockets	Any part of bottle	Cosmetic damage, possible breakage
Unfilled finish	Low spots on sealing surface	Finish surface	Leakage
Inclusions	Loose foreign material in bottle	Inside bottle	Contamination
Scabby surface	Scaly outside surface	Bottoms	Breakage
Birds-wings	Interior glass webs	Body	Glass fragments in bottle
Mold lubricant (Swab ware)	Black scale on bottle	Any part of bottle	Cosmetic damage, possible contamination
Fused glass	Glass fragments attached to bottle	Any part of bottle	Cosmetic damage, possible contamination
Out of round	Oval or elliptical shape	Body	Lack of fit with handling parts
Thin spots	Irregular glass distribution	Any part of bottle	Breakage
Dimensional problems	Bottle dimensions out of specification	Any part of bottle	Nonfunctional container
Lean tolerance	Neck leaning away from perpendicular	Neck	Misalignment in handling
Glass thickness	Variation in thickness	Any part of bottle	Breakage

bottle, so that repairs can be made. In addition, bottles usually carry an ink mark indicating the production time.

47. What types of carton materials are available?

Cartons are made from corrugated board or chipboard. Corrugated containers are generally made with two or more liners and a corrugated medium. The most popular corrugated flute types are B, C, and E. The height of the flutes and the number of flutes per running foot determine the flute type:

B flute = 50 ± 3 per foot; height = 1/8 in.,
C flute = 42 ± 3 per foot; height = 3/16 in.,
E flute = 105 ± 5 per foot; height = 3/64 in.,

B and C flutes offer good resistance to impact damage. E flute offers a higher-quality surface for printing over B or C flute. Because of the high number of flutes per foot, the print surface is smoother.

Chipboard is a nonfluted fiberboard commonly used for outer wraps of cans and half-case packs of 12 bottles each.

48. What types of liners are available?

Various outer liners are available from carton suppliers. Liner selection is guided by the printing process used and the graphic appearance desired. The following types of paper are commonly used for liners:

a. Kraft, a plain, natural, light brown paper
b. Mottle white, a bleached two-ply sheet applied to a kraft back
c. Bleached, a sheet bleached all the way through
d. Clay-coated, with either a single or a double coating

49. What is the most common carton style?

The regular slotted container (RSC) is the most common carton style. It has two minor and two major nonoverlapping flaps on the top and bottom and is held together by tabs. RSCs are erected with the top flaps folded down on the outside, so that they can be packed more easily. This style of carton is used as a shipping container by glass manufacturers.

Another style is the auto-bottom carton, sometimes used for hand packing, since it is hand-formed without the use of glue.

50. What printing processes are used to print cartons?

a. Flexography ("flexo") is a process in which a raised image on a plate is impressed into the paper to transfer the image. Each color requires a separate plate. Problems with color registration (alignment of different colors, which are printed separately) and a lack of crispness in images can occur in flexo printing.

b. Offset lithography ("litho") is a process of printing from a flat plate. The nonprinting surface of the plate (the areas where ink is not wanted) has been treated to make it water-receptive, so that it repels the oily ink. Litho printing generally offers a broader selection of colors than flexo printing and can produce crisp images without the color registration problems of flexo. It also offers more flexibility in the images that can be used.

c. Litho label printing is a process in which a solid fiber top sheet is printed and then laminated to a single-faced liner.

51. What is a basket carrier?

A basket carrier is a six-bottle carrier with a handle, made from wet-strength chipboard (usually 0.018 gauge). It is a convenient package, allowing the consumer to carry six bottles at a time. The six-pack often carries significant marketing detail, because it is the bottle package seen most on the shelf by the consumer. Basket carriers should have sufficient wet strength that they do not come apart in the grocery "well" or in ice chests.

52. What types of conveyors are common on a typical production line?

Many styles of conveyors are available for different applications in a bottling line:

a. Stainless tabletop conveyors are long-lasting but very expensive. A bottle conveyor "line lube" must be used, to reduce bottle-to-conveyor friction and drag

b. Plastic tabletop conveyors are less expensive and do not require conveyor line lube.

c. Mat-top conveyors are long-lasting but expensive. They do not require line lube.

d. Belted conveyors are good for transporting full or empty cartons short or long distances and are relatively inexpensive. Auto stop and start controls may be needed to reduce scuffing of the bottoms of cases as they are conveyed.

e. Roller conveyors may be driven or nondriven. They are used for transporting cartons on level surfaces, up slight inclines, or down steeper inclines.

f. Skate-wheel conveyors can be used for transporting cartons down a decline.

g. Bottle accumulation tables are a great help when space is restrictive. Accumulation can be achieved many ways. One common type is the bidirectional conveyor, which runs in two directions, both loading and unloading bottles as the line dictates. Accumulation tables act as surge space for the line as production speed varies.

53. What is the importance of using line lube?

Glass containers run on stainless steel or plastic conveyors. This creates a large amount of friction, especially by the stainless steel. Line lube reduces friction, drag, and noise, if a synthetic lube is used, and it can help

Figure 4.7. Uncaser removing bottles from prepacked shipping containers. (Courtesy of BridgePort Brewing Company)

keep the conveyors clean. Plastic tabletop and mat top conveyors do not require line lube. The most common lube material is made from vegetable fats with a base such as palm oil, which has good lubricity but encourages molds and microbial growth. Synthetic lubes are more expensive per gallon, but they can be used in smaller amounts and they reduce microbial growth. Carton conveyors are normally kept dry and do not require lube.

54. How are bottles delivered by the manufacturer?

Bottles come from the manufacturer either as bulk glass or in shipping containers.

a. Bulk glass. Individual bottles are stacked on a pallet, with tier sheets between layers. They are generally dispensed by a bulk glass depalletizer. A brewery that uses bulk glass will need machines for erecting cartons and basket carriers and the personnel to operate them.

Figure 4.8. Case drop packer. (Courtesy of Bridge-Port Brewing Company)

b. Shipping containers.   Bottles are packed into the brewery's cartons by the glass manufacturer, using its own machinery and personnel. A brewery that uses glass delivered in shipping containers needs only bottle-uncasing equipment (*Figure 4.7*).

55. What methods are used to pack bottles into cartons?

Many bottle case packers are available, but there are only two basic methods of packing bottles into cartons: drop packing and end loading.

a. Drop packers feed full, crowned, labeled bottles down bottle lanes above incoming empty cartons. Bottles drop through a bottle grid and are directed into a carton by grid fingers (*Figure 4.8*). Once the bottles have dropped, the case exits the machine.

b. End-loading machines both erect cartons and pack bottles. Chipboard cartons are fed from a hopper and then formed but left open on the ends. As a carton travels through the packer, it meets the bottles,

Figure 4.9. Case sealer. (Courtesy of BridgePort Brewing Company)

which are pushed into one of both of its open ends. It then travels through a sealing section and is discharged.

56. Shipping carton flaps are tabbed down on opposite corners. What method is used to open the flaps prior to sealing?

A machine called a tab slitter or tab breaker is used after the drop packer and prior to sealing. This machine cuts or breaks the two tabs so that the flaps can be closed for sealing.

57. How are the top flaps of the cases sealed?

The top flaps are closed and sealed by a case sealer (*Figure 4.9*), which consists of a case conveyor and a top-mounted belt to hold down

Figure 4.10. Floor-level palletizer. (Courtesy of BridgePort Brewing Company)

the flaps after hot-melt adhesive is applied. The cases are commonly marked with production or pull date codes during the sealing step.

58. What size choices are there in pallet boards?

Pallets are normally made of wood, although plastic pallets are also available. The pallet size depends on the case size, the number of cases per layer, and the number of layers per pallet. The number of layers may be dictated by whether cases are palletized by hand, by a floor palletizer, or by an elevated palletizer. A floor-level palletizer capable of automatically stacking cases in several different patterns up to five layers high is shown in *Figure 4.10.*

59. How are bottles and cases inspected on the line?

Automated inspection has largely replaced tedious visual inspection of bottle fill height, crowns, and full cases. There are various inspection methods, including the use of radioisotopes and harmonics. Fill height inspectors should be located as far from the filler as possible, but before the labeler, so that foam from the filling process has time to settle and a uniform height is set. Case inspectors are located after the case packer but before the case sealer.

60. What control of bottling from an economic basis is necessary?

Bottling incurs a very large part of the cost of goods in the production of beer. To be efficient, bottle shop operations should include the following controls:

a. Beer quality. Procedures for the release of beer from the finishing cellar to the filler are essential. Routine checks for desired characteristics and conformity to specifications should be made during the bottling run, and procedures should be established for separating and holding beer that is out of specification. Procedures should also be established for coding bottles, cases, and pallets with information such as production or pull dates, batch or brew numbers, and production times.

b. Labor costs. Use of packaging labor in setup, production, and cleanup should be efficient.

c. Running expenses. Beer and material losses should be controlled, down time should be minimized, and utility usage should be monitored and conserved.

d. Production cost. Cost analysis and comparisons should be constantly conducted to control the cost of materials, including labels, crowns, glass, cartons, basket carriers, and glue.

e. Maintenance. Line efficiencies should be controlled with a sound maintenance program.

61. What is included in bottle shop calculations based on a 12-oz bottle?

a. Case factor. The case factor is the amount of beer contained in a case or in 24 containers, expressed in decimal parts of a barrel. It can be determined by the following formula:

(Ounces of beer in a case) ÷ (31 gal × 128 oz/gal)
= (Ounces of beer in a case) ÷ 3,968 oz
= (Amount of beer in cases, expressed in decimal parts of a barrel)

b. Amount of beer (in barrels) received in case stock

(Number of cases received in stock) × (Case factor)

c. Beer loss (in barrels)

(Amount of beer taken from finishing cellar stock or beer meter and filled off) − (Amount of beer received in case stock)

d. Percentage of beer loss

(Barrels lost) ÷ (Barrels filled off) × 100

e. Beer loss charges

(Barrels lost) × (Cost of beer in bottles per barrel)

f. Cost of one empty beer bottle

(Cost per gross) ÷ 144

g. Bottle loss charges

(Cost of one bottle) × (Number of bottles lost)

h. Percentage of bottles lost

(Number of bottles lost) × 100 ÷ (Number of bottles received)

i. Value of beer in one case

(Cost per barrel) × (Case factor)

j. Cost of crowns per 24-bottle case

(Cost per gross) ÷ 6

k. Cost of labels per 24-bottle case

(Cost per 1,000) × 24 ÷ 1,000

l. Loss charges per case

(Total loss charges)
    ÷ (Total number of cases filled off and received in stock)

m. Labor charges per case

(Total labor charges)
    ÷ (Total number of cases filled off and received in stock)

CHAPTER 5

Laboratory Methods and Instruments

Candace E. Wallin

University of California, Davis

1. What is a hydrometer?

A hydrometer is an instrument used to determine the specific gravity of liquids. It is allowed to float in a liquid, and the depth to which it sinks is measured. The hydrometer, when floating, displaces its own weight of liquid and, therefore, sinks deeper in a light liquid than in a heavy liquid. Hydrometers are usually made of glass and consist of a small bulb filled with a heavy material near the bottom, which serves to keep the instrument in an upright position when floating, and a larger bulb filled with air, which keeps the instrument afloat. The bulbs are connected by a long, slender stem on which a scale is marked, measuring the specific gravity (or degrees Plato) of the liquid. Hydrometers often have a temperature compensation scale on the large bulb to provide accurate specific gravity readings at temperatures other than 68°F (20°C).

2. What are degrees Balling and degrees Plato?

The extract tables are named after Karl J. N. Balling and Dr. Plato, the German scientists who established them in the nineteenth century. A solution of 1° Balling (1° Plato) has the same specific gravity as a cane sugar solution consisting of 1 lb of cane sugar dissolved in enough water to make 100 lb of solution. In other words, 1° Balling (1° Plato) is 1% by weight. The Balling tables (produced in 1843) are based on sugar solutions at 63.5°F (17.5°C). The Plato tables, based on sugar solutions at 68°F (20°C), were devised to correct slight errors in the Balling tables. The tables produced by Plato remain the standard in the brewing industry. The differences in the tables are very small, and brewers use the terms *degrees Balling* and *degrees Plato* interchangeably.

3. What is specific gravity?

The specific gravity of a substance (solid or liquid) is the ratio of the weight of any volume of the substance to the weight of an equal volume of water. Calculation of specific gravity is illustrated in the following example:

> Weight of flask filled with wort: 25 oz
> Weight of same flask filled with water: 22 oz
> Weight of empty flask: 10 oz
> Weight of wort: 25 oz – 10 oz = 15 oz
> Weight of water: 22 oz – 10 oz = 12 oz
> Specific gravity of wort: 15 oz/12 oz = 1.25

Specific gravity can be also be determined using a hydrometer. *Table 5.1* relates specific gravity to degrees Plato. In the metric system of measurement, specific gravity is equal to density.

4. What is an extract?

Extracts are substances that will dissolve into solution. In brewing, water is used to extract sugars from the various brewing ingredients. For example, for every 100 lb of malt used, 80 lb of starches, proteins, and other materials may be extracted during brewing. Husks and spent grains remaining in the wort separation vessel account for the remaining 20 lb of malt.

The maltster analyzes each lot of malt and sends the results of pertinent analyses, including the amount of extract available in coarse and/or fine grinds, for every lot of malt delivered. Since a laboratory fine grind yields an almost unattainable amount of extract, most brewers use a laboratory coarse grind in calculating their material bills and brewhouse efficiencies.

Any extract in the wort in excess of that obtained from malt is derived from other materials used in brewing.

5. How is the Plato reading related to the amount of extract per barrel?

Assuming that the specific gravity of finished beer is very close to that of water, using the weight of water (259 lb at 4°C [approximately

Table 5.1. Specific gravity and degrees Plato of sugar solutions or percent extract by weight

Specific gravity at 20°C/20°C	Grams extract in 100 g of solution	Specific gravity at 20°C/20°C	Grams extract in 100 g of solution	Specific gravity at 20°C/20°C	Grams extract in 100 g of solution	Specific gravity at 20°C/20°C	Grams extract in 100 g of solution	Specific gravity at 20°C/20°C	Grams extract in 100 g of solution
1.00000	0.000	1.00250	0.642	1.00500	1.283	1.00750	1.923	1.01000	2.560
05	0.013	55	0.655	05	1.296	55	1.935	05	2.572
10	0.026	60	0.668	10	1.308	60	1.948	10	2.585
15	0.039	65	0.680	15	1.321	65	1.961	15	2.598
20	0.052	70	0.693	20	1.334	70	1.973	20	2.610
25	0.064	75	0.706	25	1.347	75	1.986	25	2.623
30	0.077	80	0.719	30	1.360	80	1.999	30	2.636
35	0.090	85	0.732	35	1.372	85	2.012	35	2.649
40	0.103	90	0.745	40	1.385	90	2.025	40	2.661
45	0.116	95	0.757	45	1.398	95	2.038	45	2.674
1.00050	0.129	1.00300	0.770	1.00550	1.411	1.00800	2.051	1.01050	2.687
55	0.141	05	0.783	55	1.424	05	2.065	55	2.699
60	0.154	10	0.796	60	1.437	10	2.078	60	2.712
65	0.167	15	0.808	65	1.450	15	2.090	65	2.725
70	0.180	20	0.821	70	1.462	20	2.102	70	2.738
75	0.193	25	0.834	75	1.475	25	2.114	75	2.750
80	0.206	30	0.847	80	1.488	30	2.127	80	2.763
85	0.219	35	0.859	85	1.501	35	2.139	85	2.776
90	0.231	40	0.872	90	1.514	40	2.152	90	2.788
95	0.244	45	0.885	95	1.526	45	2.165	95	2.801
1.00100	0.257	1.00350	0.898	1.00600	1.539	1.00850	2.178	1.01100	2.814
05	0.270	55	0.911	05	1.552	55	2.191	05	2.826
10	0.283	60	0.924	10	1.565	60	2.203	10	2.839
15	0.296	65	0.937	15	1.578	65	2.216	15	2.852
20	0.309	70	0.949	20	1.590	70	2.229	20	2.864
25	0.321	75	0.962	25	1.603	75	2.241	25	2.877
30	0.334	80	0.975	30	1.616	80	2.254	30	2.890
35	0.347	85	0.988	35	1.629	85	2.267	35	2.903
40	0.360	90	1.001	40	1.641	90	2.280	40	2.915
45	0.373	95	1.014	45	1.654	95	2.292	45	2.928
1.00150	0.386	1.00400	1.026	1.00650	1.667	1.00900	2.305	1.01150	2.940
55	0.398	05	1.039	55	1.680	05	2.317	55	2.953
60	0.411	10	1.052	60	1.693	10	2.330	60	2.966
65	0.424	15	1.065	65	1.705	15	2.343	65	2.979
70	0.437	20	1.078	70	1.718	20	2.356	70	2.991
75	0.450	25	1.090	75	1.731	25	2.369	75	3.004
80	0.463	30	1.103	80	1.744	30	2.381	80	3.017
85	0.476	35	1.116	85	1.757	35	2.394	85	3.029
90	0.488	40	1.129	90	1.769	40	2.407	90	3.042
95	0.501	45	1.142	95	1.782	45	2.419	95	3.055
1.00200	0.514	1.00450	1.155	1.00700	1.795	1.00950	2.432	1.01200	3.067
05	0.527	55	1.168	05	1.807	55	2.445	05	3.080
10	0.540	60	1.180	10	1.820	60	2.458	10	3.093
15	0.552	65	1.193	15	1.833	65	2.470	15	3.105
20	0.565	70	1.206	20	1.846	70	2.483	20	3.118
25	0.578	75	1.219	25	1.859	75	2.496	25	3.131
30	0.591	80	1.232	30	1.872	80	2.508	30	3.143
35	0.604	85	1.244	35	1.884	85	2.521	35	3.156
40	0.616	90	1.257	40	1.897	90	2.534	40	3.169
45	0.629	95	1.270	45	1.910	95	2.547	45	3.181

(continued on next page)

Reprinted, by permission, from American Society of Brewing Chemists, 2004, *Methods of Analysis,* 9th ed., ASBC, St. Paul, Minn.

Table 5.1 (*continued*). Specific gravity and degrees Plato of sugar solutions or percent extract by weight

Specific gravity at 20°C/20°C	Grams extract in 100 g of solution	Specific gravity at 20°C/20°C	Grams extract in 100 g of solution	Specific gravity at 20°C/20°C	Grams extract in 100 g of solution	Specific gravity at 20°C/20°C	Grams extract in 100 g of solution	Specific gravity at 20°C/20°C	Grams extract in 100 g of solutions
1.01250	3.194	1.01500	3.826	1.01750	4.454	1.02000	5.080	1.02250	5.704
55	3.207	05	3.838	55	4.467	05	5.093	55	5.716
60	3.219	10	3.851	60	4.479	10	5.106	60	5.729
65	3.232	15	3.863	65	4.492	15	5.118	65	5.741
70	3.245	20	3.876	70	4.505	20	5.130	70	5.754
75	3.257	25	3.888	75	4.517	25	5.143	75	5.766
80	3.270	30	3.901	80	4.529	30	5.155	80	5.779
85	3.282	35	3.914	85	4.542	35	5.168	85	5.791
90	3.295	40	3.926	90	4.555	40	5.180	90	5.803
95	3.308	45	3.939	95	4.567	45	5.193	95	5.816
1.01300	3.321	1.01550	3.951	1.01800	4.580	1.02050	5.205	1.02300	5.828
05	3.333	55	3.964	05	4.592	55	5.218	05	5.841
10	3.346	60	3.977	10	4.605	60	5.230	10	5.853
15	3.358	65	3.989	15	4.617	65	5.243	15	5.865
20	3.371	70	4.002	20	4.630	70	5.255	20	5.878
25	3.384	75	4.014	25	4.642	75	5.268	25	5.890
30	3.396	80	4.027	30	4.655	80	5.280	30	5.903
35	3.409	85	4.039	35	4.668	85	5.293	35	5.915
40	3.421	90	4.052	40	4.680	90	5.305	40	5.928
45	3.434	95	4.065	45	4.692	95	5.318	45	5.940
1.01350	3.447	1.01600	4.077	1.01850	4.705	1.02100	5.330	1.02350	5.952
55	3.459	05	4.090	55	4.718	05	5.343	55	5.965
60	3.472	10	4.102	60	4.730	10	5.355	60	5.977
65	3.485	15	4.115	65	4.743	15	5.367	65	5.990
70	3.497	20	4.128	70	4.755	20	5.380	70	6.002
75	3.510	25	4.140	75	4.768	25	5.392	75	6.015
80	3.523	30	4.153	80	4.780	30	5.405	80	6.027
85	3.535	35	4.165	85	4.792	35	5.418	85	6.039
90	3.548	40	4.178	90	4.805	40	5.430	90	6.052
95	3.561	45	4.190	95	4.818	45	5.443	95	6.064
1.01400	3.573	1.01650	4.203	1.01900	4.830	1.02150	5.455	1.02400	6.077
05	3.586	55	4.216	05	4.843	55	5.467	05	6.089
10	3.598	60	4.228	10	4.855	60	5.480	10	6.101
15	3.611	65	4.241	15	4.868	65	5.492	15	6.114
20	3.624	70	4.253	20	4.880	70	5.505	20	6.126
25	3.636	75	4.266	25	4.893	75	5.517	25	6.139
30	3.649	80	4.278	30	4.905	80	5.530	30	6.151
35	3.662	85	4.291	35	4.918	85	5.542	35	6.163
40	3.674	90	4.304	40	4.930	90	5.555	40	6.176
45	3.687	95	4.316	45	4.943	95	5.567	45	6.188
1.01450	3.699	1.01700	4.329	1.01950	4.955	1.02200	5.580	1.02450	6.200
55	3.712	05	4.341	55	4.968	05	5.592	55	6.213
60	3.725	10	4.354	60	4.980	10	5.605	60	6.225
65	3.737	15	4.366	65	4.993	15	5.617	65	6.238
70	3.750	20	4.379	70	5.006	20	5.629	70	6.250
75	3.762	25	4.391	75	5.018	25	5.642	75	6.263
80	3.775	30	4.404	80	5.030	30	5.654	80	6.275
85	3.788	35	4.417	85	5.043	35	5.667	85	6.287
90	3.800	40	4.429	90	5.055	40	5.679	90	6.300
95	3.813	45	4.442	95	5.068	45	5.692	95	6.312

(*continued on next page*)

Table 5.1 (*continued*). Specific gravity and degrees Plato of sugar solutions or per-cent extract by weight

Specific gravity at 20°C/20°C	Grams extract in 100 g of solution	Specific gravity at 20°C/20°C	Grams extract in 100 g of solution	Specific gravity at 20°C/20°C	Grams extract in 100 g of solution	Specific gravity at 20°C/20°C	Grams extract in 100 g of solution	Specific gravity at 20°C/20°C	Grams extract in 100 g of solution
1.02500	6.325	1.02750	6.943	1.03000	7.558	1.03250	8.171	1.03500	8.781
05	6.337	55	6.955	05	7.570	55	8.183	05	8.793
10	6.350	60	6.967	10	7.583	60	8.195	10	8.805
15	6.362	65	6.979	15	7.595	65	8.207	15	8.817
20	6.374	70	6.992	20	7.607	70	8.220	20	8.830
25	6.387	75	7.004	25	7.619	75	8.232	25	8.842
30	6.399	80	7.017	30	7.632	80	8.244	30	8.854
35	6.411	85	7.029	35	7.644	85	8.256	35	8.866
40	6.424	90	7.041	40	7.656	90	8.269	40	8.878
45	6.436	95	7.053	45	7.668	95	8.281	45	8.890
1.02550	6.449	1.02800	7.066	1.03050	7.681	1.03300	8.293	1.03550	8.902
55	6.461	05	7.078	55	7.693	05	8.305	55	8.915
60	6.473	10	7.091	60	7.705	10	8.317	60	8.927
65	6.485	15	7.103	65	7.717	15	8.330	65	8.939
70	6.498	20	7.115	70	7.730	20	8.342	70	8.951
75	6.510	25	7.127	75	7.742	25	8.354	75	8.963
80	6.523	30	7.140	80	7.754	30	8.366	80	8.975
85	6.535	35	7.152	85	7.767	35	8.378	85	8.988
90	6.547	40	7.164	90	7.779	40	8.391	90	9.000
95	6.560	45	7.177	95	7.791	45	8.403	95	9.012
1.02600	6.572	1.02850	7.189	1.03100	7.803	1.03350	8.415	1.03600	9.024
05	6.584	55	7.201	05	7.816	55	8.427	05	9.036
10	6.597	60	7.214	10	7.828	60	8.439	10	9.048
15	6.609	65	7.226	15	7.840	65	8.452	15	9.060
20	6.621	70	7.238	20	7.853	70	8.464	20	9.073
25	6.634	75	7.251	25	7.865	75	8.476	25	9.085
30	6.646	80	7.263	30	7.877	80	8.488	30	9.097
35	6.659	85	7.275	35	7.889	85	8.500	35	9.109
40	6.671	90	7.287	40	7.901	90	8.513	40	9.121
45	6.683	95	7.300	45	7.914	95	8.525	45	9.133
1.02650	6.696	1.02900	7.312	1.03150	7.926	1.03400	8.537	1.03650	9.145
55	6.708	05	7.324	55	7.938	05	8.549	55	9.158
60	6.720	10	7.337	60	7.950	10	8.561	60	9.170
65	6.733	15	7.349	65	7.963	15	8.574	65	9.182
70	6.745	20	7.361	70	7.975	20	8.586	70	9.194
75	6.757	25	7.374	75	7.987	25	8.598	75	9.206
80	6.770	30	7.386	80	8.000	30	8.610	80	9.218
85	6.782	35	7.398	85	8.012	35	8.622	85	9.230
90	6.794	40	7.411	90	8.024	40	8.634	90	9.243
95	6.807	45	7.423	95	8.036	45	8.647	95	9.255
1.02700	6.819	1.02950	7.435	1.03200	8.048	1.03450	8.659	1.03700	9.267
05	6.831	55	7.447	05	8.061	55	8.671	05	9.279
10	6.844	60	7.460	10	8.073	60	8.683	10	9.291
15	6.856	65	7.472	15	8.085	65	8.695	15	9.303
20	6.868	70	7.484	20	8.098	70	8.708	20	9.316
25	6.881	75	7.497	25	8.110	75	8.720	25	9.328
30	6.893	80	7.509	30	8.122	80	8.732	30	9.340
35	6.905	85	7.521	35	8.134	85	8.744	35	9.352
40	6.918	90	7.533	40	8.146	90	8.756	40	9.364
45	6.930	95	7.546	45	8.159	95	8.768	45	9.376

(*continued on next page*)

Table 5.1 (*continued*). Specific gravity and degrees Plato of sugar solutions or percent extract by weight

Specific gravity at 20°C/20°C	Grams extract in 100 g of solution	Specific gravity at 20°C/20°C	Grams extract in 100 g of solution	Specific gravity at 20°C/20°C	Grams extract in 100 g of solution	Specific gravity at 20°C/20°C	Grams extract in 100 g of solution	Specific gravity at 20°C/20°C	Grams extract in 100 g of solution
1.03750	9.388	1.04000	9.993	1.04250	10.596	1.04500	11.195	1.04750	11.792
55	9.400	05	10.005	55	10.608	05	11.207	55	11.804
60	9.413	10	10.017	60	10.620	10	11.219	60	11.816
65	9.425	15	10.030	65	10.632	15	11.231	65	11.828
70	9.437	20	10.042	70	10.644	20	11.243	70	11.840
75	9.449	25	10.054	75	10.656	25	11.255	75	11.852
80	9.461	30	10.066	80	10.668	30	11.267	80	11.864
85	9.473	35	10.078	85	10.680	35	11.279	85	11.876
90	9.485	40	10.090	90	10.692	40	11.291	90	11.888
95	9.498	45	10.102	95	10.704	45	11.303	95	11.900
1.03800	9.509	1.04050	10.114	1.04300	10.716	1.04550	11.315	1.04800	11.912
05	9.522	55	10.126	05	10.728	55	11.327	05	11.923
10	9.534	60	10.138	10	10.740	60	11.339	10	11.935
15	9.546	65	10.150	15	10.752	65	11.351	15	11.947
20	9.558	70	10.162	20	10.764	70	11.363	20	11.959
25	9.570	75	10.174	25	10.776	75	11.375	25	11.971
30	9.582	80	10.186	30	10.788	80	11.387	30	11.983
35	9.594	85	10.198	35	10.800	85	11.399	35	11.995
40	9.606	90	10.210	40	10.812	90	11.411	40	12.007
45	9.618	95	10.223	45	10.824	95	11.423	45	12.019
1.03850	9.631	1.04100	10.234	1.04350	10.836	1.04600	11.435	1.04850	12.031
55	9.643	05	10.246	55	10.848	05	11.446	55	12.042
60	9.655	10	10.259	60	10.860	10	11.458	60	12.054
65	9.667	15	10.271	65	10.872	15	11.470	65	12.066
70	9.679	20	10.283	70	10.884	20	11.482	70	12.078
75	9.691	25	10.295	75	10.896	25	11.494	75	12.090
80	9.703	30	10.307	80	10.908	30	11.506	80	12.102
85	9.715	35	10.319	85	10.920	35	11.518	85	12.114
90	9.727	40	10.331	90	10.932	40	11.530	90	12.126
95	9.740	45	10.343	95	10.944	45	11.542	95	12.138
1.03900	9.751	1.04150	10.355	1.04400	10.956	1.04650	11.554	1.04900	12.150
05	9.764	55	10.367	05	10.968	55	11.566	05	12.162
10	9.776	60	10.379	10	10.980	60	11.578	10	12.173
15	9.788	65	10.391	15	10.992	65	11.590	15	12.185
20	9.800	70	10.403	20	11.004	70	11.602	20	12.197
25	9.812	75	10.415	25	11.016	75	11.614	25	12.209
30	9.824	80	10.427	30	11.027	80	11.626	30	12.221
35	9.836	85	10.439	35	11.039	85	11.638	35	12.233
40	9.848	90	10.451	40	11.051	90	11.650	40	12.245
45	9.860	95	10.463	45	11.063	95	11.661	45	12.256
1.03950	9.873	1.04200	10.475	1.04450	11.075	1.04700	11.673	1.04950	12.268
55	9.885	05	10.487	55	11.087	05	11.685	55	12.280
60	9.897	10	10.499	60	11.100	10	11.697	60	12.292
65	9.909	15	10.511	65	11.112	15	11.709	65	12.304
70	9.921	20	10.523	70	11.123	20	11.721	70	12.316
75	9.933	25	10.536	75	11.135	25	11.733	75	12.328
80	9.945	30	10.548	80	11.147	30	11.745	80	12.340
85	9.957	35	10.559	85	11.159	35	11.757	85	12.351
90	9.969	40	10.571	90	11.171	40	11.768	90	12.363
95	9.981	45	10.584	95	11.183	45	11.780	95	12.375

(*continued on next page*)

Table 5.1 (*continued*). Specific gravity and degrees Plato of sugar solutions or percent extract by weight

Specific gravity at 20°C/20°C	Grams extract in 100 g of solution	Specific gravity at 20°C/20°C	Grams extract in 100 g of solution	Specific gravity at 20°C/20°C	Grams extract in 100 g of solution	Specific gravity at 20°C/20°C	Grams extract in 100 g of solution	Specific gravity at 20°C/20°C	Grams extract in 100 g of solution
1.05000	12.387	1.05250	12.979	1.05500	13.569	1.05750	14.156	1.06000	14.741
05	12.399	55	12.991	05	13.581	55	14.168	05	14.752
10	12.411	60	13.003	10	13.593	60	14.179	10	14.764
15	12.423	65	13.015	15	13.604	65	14.191	15	14.776
20	12.435	70	13.027	20	13.616	70	14.203	20	14.787
25	12.447	75	13.039	25	13.628	75	14.215	25	14.799
30	12.458	80	13.050	30	13.640	80	14.226	30	14.811
35	12.470	85	13.062	35	13.651	85	14.238	35	14.822
40	12.482	90	13.074	40	13.663	90	14.250	40	14.834
45	12.494	95	13.086	45	13.675	95	14.261	45	14.846
1.05050	12.506	1.05300	13.098	1.05550	13.687	1.05800	14.273	1.06050	14.857
55	12.518	05	13.109	55	13.698	05	14.285	55	14.869
60	12.530	10	13.121	60	13.710	10	14.297	60	14.881
65	12.542	15	13.133	65	13.722	15	14.308	65	14.892
70	12.553	20	13.145	70	13.734	20	14.320	70	14.904
75	12.565	25	13.157	75	13.746	25	14.332	75	14.916
80	12.577	30	13.168	80	13.757	30	14.343	80	14.927
85	12.589	35	13.180	85	13.769	35	14.355	85	14.939
90	12.601	40	13.192	90	13.781	40	14.367	90	14.950
95	12.613	45	13.204	95	13.792	45	14.379	95	14.962
1.05100	12.624	1.05350	13.215	1.05600	13.804	1.05850	14.390	1.06100	14.974
05	12.636	55	13.227	05	13.816	55	14.402	05	14.986
10	12.648	60	13.239	10	13.828	60	14.414	10	14.997
15	12.660	65	13.251	15	13.839	65	14.425	15	15.009
20	12.672	70	13.263	20	13.851	70	14.437	20	15.020
25	12.684	75	13.274	25	13.863	75	14.449	25	15.032
30	12.695	80	13.286	30	13.875	80	14.460	30	15.044
35	12.707	85	13.298	35	13.886	85	14.472	35	15.055
40	12.719	90	13.310	40	13.898	90	14.484	40	15.067
45	12.731	95	13.322	45	13.910	95	14.495	45	15.079
1.05150	12.743	1.054000	13.333	1.05650	13.921	1.05900	14.507	1.06150	15.090
55	12.755	05	13.345	55	13.933	05	14.519	55	15.102
60	12.767	10	13.357	60	13.945	10	14.531	60	15.114
65	12.778	15	13.369	65	13.957	15	14.542	65	15.125
70	12.790	20	13.380	70	13.968	20	14.554	70	15.137
75	12.802	25	13.392	75	13.980	25	14.565	75	15.148
80	12.814	30	13.404	80	13.992	30	14.577	80	15.160
85	12.826	35	13.416	85	14.004	35	14.589	85	15.172
90	12.838	40	13.428	90	14.015	40	14.601	90	15.183
95	12.849	45	13.439	95	14.027	45	14.612	95	15.195
1.05200	12.861	1.05450	13.451	1.05700	14.039	1.05950	14.624	1.06200	15.207
05	12.873	55	13.463	05	14.051	55	14.636	05	15.218
10	12.885	60	13.475	10	14.062	60	14.647	10	15.230
15	12.897	65	13.487	15	14.074	65	14.659	15	15.241
20	12.909	70	13.499	20	14.086	70	14.671	20	15.253
25	12.920	75	13.510	25	14.097	75	14.682	25	15.265
30	12.932	80	13.522	30	14.109	80	14.694	30	15.276
35	12.944	85	13.534	35	14.121	85	14.706	35	15.288
40	12.956	90	13.546	40	14.133	90	14.717	40	15.300
45	12.968	95	13.557	45	14.144	95	14.729	45	15.311

(*continued on next page*)

Table 5.1 (*continued*). Specific gravity and degrees Plato of sugar solutions or percent extract by weight

Specific gravity at 20°C/20°C	Grams extract in 100 g of solution	Specific gravity at 20°C/20°C	Grams extract in 100 g of solution	Specific gravity at 20°C/20°C	Grams extract in 100 g of solution	Specific gravity at 20°C/20°C	Grams extract in 100 g of solution	Specific gravity at 20°C/20°C	Grams extract in 100 g of solution
1.06250	15.323	1.06500	15.903	1.06750	16.480	1.07000	17.055	1.07250	17.627
55	15.334	05	15.914	55	16.491	05	17.066	55	17.639
60	15.346	10	15.926	60	16.503	10	17.078	60	17.650
65	15.358	15	15.938	65	16.514	15	17.089	65	17.661
70	15.369	20	15.949	70	16.526	20	17.101	70	17.673
75	15.381	25	15.961	75	16.537	25	17.112	75	17.684
80	15.393	30	15.972	80	16.549	30	17.123	80	17.696
85	15.404	35	15.984	85	16.561	35	17.135	85	17.707
90	15.416	40	15.995	90	16.572	40	17.146	90	17.719
95	15.427	45	16.007	95	16.583	45	17.158	95	17.730
1.06300	15.439	1.06550	16.019	1.06800	16.595	1.07050	17.169	1.07300	17.741
05	15.451	55	16.030	05	16.606	55	17.181	05	17.753
10	15.462	60	16.041	10	16.618	60	17.192	10	17.764
15	15.474	65	16.053	15	16.630	65	17.204	15	17.776
20	15.486	70	16.065	20	16.641	70	17.215	20	17.787
25	15.497	75	16.076	25	16.652	75	17.227	25	17.799
30	15.509	80	16.088	30	16.664	80	17.238	30	17.810
35	15.520	85	16.099	35	16.676	85	17.250	35	17.821
40	15.532	90	16.111	40	16.687	90	17.261	40	17.833
45	15.544	95	16.122	45	16.699	95	17.272	45	17.844
1.06350	15.555	1.06600	16.134	1.06850	16.710	1.07100	17.284	1.07350	17.856
55	15.567	05	16.145	55	16.722	05	17.295	55	17.867
60	15.578	10	16.157	60	16.733	10	17.307	60	17.878
65	15.590	15	16.169	65	16.744	15	17.318	65	17.890
70	15.602	20	16.180	70	16.756	20	17.330	70	17.901
75	15.613	25	16.191	75	16.768	25	17.341	75	17.913
80	15.625	30	16.203	80	16.779	30	17.353	80	17.924
85	15.637	35	16.215	85	16.791	35	17.364	85	17.935
90	15.648	40	16.226	90	16.802	40	17.375	90	17.947
95	15.660	45	16.238	95	16.813	45	17.387	95	17.958
1.06400	15.671	1.06650	16.249	1.06900	16.825	1.07150	17.398	1.07400	17.970
05	15.683	55	16.261	05	16.836	55	17.410	05	17.981
10	15.694	60	16.272	10	16.848	60	17.421	10	17.992
15	15.706	65	16.284	15	16.859	65	17.433	15	18.004
20	15.717	70	16.295	20	16.871	70	17.444	20	18.015
25	15.729	75	16.307	25	16.882	75	17.456	25	18.027
30	15.741	80	16.319	30	16.894	80	17.467	30	18.038
35	15.752	85	16.330	35	16.905	85	17.479	35	18.049
40	15.764	90	16.341	40	16.917	90	17.490	40	18.061
45	15.776	95	16.353	45	16.928	95	17.501	45	18.072
1.06450	15.787	1.06700	16.365	1.06950	16.940	1.07200	17.513	1.07450	18.084
55	15.799	05	16.376	55	16.951	05	17.524	55	18.095
60	15.810	10	16.388	60	16.963	10	17.536	60	18.106
65	15.822	15	16.399	65	16.974	15	17.547	65	18.118
70	15.833	20	16.411	70	16.986	20	17.559	70	18.129
75	15.845	25	16.422	75	16.997	25	17.570	75	18.140
80	15.857	30	16.434	80	17.009	30	17.581	80	18.152
85	15.868	35	16.445	85	17.020	35	17.593	85	18.163
90	15.880	40	16.457	90	17.032	40	17.604	90	18.175
95	15.891	45	16.468	95	17.043	45	17.616	95	18.186

(*continued on next page*)

Table 5.1 (*continued*). Specific gravity and degrees Plato of sugar solutions or percent extract by weight

Specific gravity at 20°C/20°C	Grams extract in 100 g of solution	Specific gravity at 20°C/20°C	Grams extract in 100 g of solution	Specific gravity at 20°C/20°C	Grams extract in 100 g of solution	Specific gravity at 20°C/20°C	Grams extract in 100 g of solution
1.07500	18.197	1.07750	18.765	1.08000	19.331	1.08250	19.894
05	18.209	55	18.777	05	19.342	55	19.905
10	18.220	60	18.788	10	19.353	60	19.917
15	18.232	65	18.799	15	19.365	65	19.928
20	18.243	70	18.810	20	19.376	70	19.939
25	18.254	75	18.822	25	19.387	75	19.950
30	18.266	80	18.833	30	19.399	80	19.961
35	18.277	85	18.845	35	19.410	85	19.973
40	18.288	90	18.856	40	19.421	90	19.984
45	18.300	95	18.867	45	19.432	95	19.995
1.07550	18.311	1.07800	18.878	1.08050	19.444	1.08300	20.007
55	18.323	05	18.890	55	19.455		
60	18.334	10	18.901	60	19.466		
65	18.345	15	18.912	65	19.478		
70	18.356	20	18.924	70	19.489		
75	18.368	25	18.935	75	19.500		
80	18.379	30	18.947	80	19.511		
85	18.391	35	18.958	85	19.523		
90	18.402	40	18.969	90	19.534		
95	18.413	45	18.980	95	19.545		
1.07600	18.425	1.07850	18.992	1.08100	19.556		
05	18.436	55	19.003	05	19.567		
10	18.447	60	19.015	10	19.579		
15	18.459	65	19.026	15	19.590		
20	18.470	70	19.037	20	19.601		
25	18.482	75	19.048	25	19.613		
30	18.493	80	19.060	30	19.624		
35	18.504	85	19.071	35	19.635		
40	18.516	90	19.082	40	19.646		
45	18.527	95	19.094	45	19.658		
1.07650	18.538	1.07900	19.105	1.08150	19.669		
55	18.550	05	19.116	55	19.680		
60	18.561	10	19.127	60	19.692		
65	18.572	15	19.139	65	19.703		
70	18.584	20	19.150	70	19.714		
75	18.595	25	19.161	75	19.725		
80	18.607	30	19.173	80	19.737		
85	18.618	35	19.184	85	19.748		
90	18.629	40	19.195	90	19.759		
95	18.641	45	19.207	95	19.770		
1.07700	18.652	1.07950	19.218	1.08200	19.782		
05	18.663	55	19.229	05	19.793		
10	18.675	60	19.241	10	19.804		
15	18.686	65	19.252	15	19.815		
20	18.697	70	19.263	20	19.827		
25	18.709	75	19.274	25	19.838		
30	18.720	80	19.286	30	19.849		
35	18.731	85	19.297	35	19.860		
40	18.742	90	19.308	40	19.872		
45	18.754	95	19.320	45	19.883		

39°F]) we can use the following formula to convert a Plato figure into pounds of extract per barrel.

$$\frac{(259 + \text{degrees Plato}) \times (\text{degrees Plato})}{100} = \text{Extract per barrel (in pounds)}$$

Example.   For a 12° Plato wort, the amount of extract per barrel is

$$\frac{(259 + 12) \times 12}{100} = \frac{271 \times 12}{100} = \frac{3252}{100} = 32.52 \text{ lb of extract per barrel}$$

The amount of extract can also be determined from a table published by Schwarz Laboratories in 1946 (*Table 5.2*). The values in this table are not exact, but they are close enough for most practical purposes. For example, the table lists the extract per barrel for a 12.0°P wort as 32.53 lb, while the calculated value is 32.52 lb.

Note: One barrel of water (31 gal) weighs approximately 259 lb (more precisely, 258.7 lb) at 39.2°F (4.0°C).

6. What is yield?

Yield is the amount of dry extract obtained from the material used.

7. What is yield on an extract basis?

Yield calculated on an extract basis can be derived from empirical observations and data for several brews. Yield on an extract basis is the amount of extract (percentage) that can be expected per pound of material. *Table 5.3* provides some extract yield values for materials commonly used in brewing.

For example, for malt with an average yield in the brewery of 78% for each 100 lb of malt used, only 78 lb of extract is produced. This extraction efficiency is a function of the brewhouse performance from milling through wort separation and is used in brewing material bill calculations.

Since beer cannot be brewed without malt, the extract yield of the malt is usually the most important factor in determining the yield and cost of each brew. Any extract in the wort in excess of the extract yield from malt must necessarily come from other materials. Therefore, total extract yield is the sum total of extract obtained from all of the materials used.

8. What are laboratory yield and brewing yield?

a. Laboratory yield is the extract derived from the raw material (malt, corn, rice, etc.) following precise laboratory procedures. The ma-

Table 5.2. Pounds of extract per 31-gal barrel

°Plato	Extract (lb.)	°Plato	Extract (lb.)	°Plato	Extract (lb.)	°Plato	Extract (lb.)
		6.0	15.85	12.0	32.45	18.0	49.88
0.1	0.26	6.1	16.12	12.1	32.73	18.1	50.18
0.2	0.52	6.2	16.39	12.2	33.02	18.2	50.47
0.3	0.77	6.3	16.66	12.3	33.31	18.3	50.77
0.4	1.03	6.4	16.93	12.4	33.59	18.4	51.07
0.5	1.29	6.5	17.20	12.5	33.88	18.5	51.37
0.6	1.55	6.6	17.47	12.6	34.16	18.6	51.67
0.7	1.81	6.7	17.74	12.7	34.44	18.7	51.97
0.8	2.07	6.8	18.01	12.8	34.73	18.8	52.27
0.9	2.33	6.9	18.29	12.9	35.01	18.9	52.57
1.0	2.59	7.0	18.56	13.0	35.30	19.0	52.87
1.1	2.85	7.1	18.83	13.1	35.59	19.1	53.17
1.2	3.11	7.2	19.10	13.2	35.87	19.2	53.47
1.3	3.37	7.3	19.38	13.3	36.15	19.3	53.77
1.4	3.63	7.4	19.65	13.4	36.45	19.4	54.07
1.5	3.89	7.5	19.92	13.5	36.73	19.5	54.37
1.6	4.15	7.6	20.20	13.6	37.02	19.6	54.67
1.7	4.41	7.7	20.47	13.7	37.31	19.7	54.97
1.8	4.68	7.8	20.75	13.8	37.59	19.8	55.27
1.9	4.94	7.9	21.02	13.9	37.88	19.9	55.58
2.0	5.20	8.0	21.29	14.0	38.17	20.0	55.88
2.1	5.46	8.1	21.57	14.1	38.46	20.1	56.18
2.2	5.72	8.2	21.84	14.2	38.75	20.2	56.48
2.3	5.99	8.3	22.12	14.3	39.04	20.3	56.79
2.4	6.25	8.4	22.40	14.4	39.32	20.4	57.09
2.5	6.51	8.5	22.67	14.5	39.61	20.5	57.39
2.6	6.78	8.6	22.95	14.6	39.92	20.6	57.70
2.7	7.04	8.7	23.22	14.7	40.19	20.7	58.00
2.8	7.30	8.8	23.50	14.8	40.48	20.8	58.30
2.9	7.57	8.9	23.77	14.9	40.77	20.9	58.61
3.0	7.83	9.0	24.05	15.0	41.06	21.0	58.91
3.1	8.09	9.1	24.33	15.1	41.35	21.1	59.22
3.2	8.36	9.2	24.61	15.2	41.64	21.2	59.52
3.3	8.62	9.3	24.88	15.3	41.93	21.3	59.83
3.4	8.89	9.4	25.16	15.4	42.22	21.4	60.14
3.5	9.15	9.5	25.44	15.5	42.52	21.5	60.44
3.6	9.42	9.6	25.72	15.6	42.81	21.6	60.75
3.7	9.68	9.7	25.99	15.7	43.10	21.7	61.05
3.8	9.95	9.8	26.27	15.8	43.39	21.8	61.36
3.9	10.21	9.9	26.55	15.9	43.68	21.9	61.67
4.0	10.48	10.0	26.83	16.0	43.97	22.0	61.97
4.1	10.75	10.1	27.11	16.1	44.27	22.1	62.28
4.2	11.01	10.2	27.38	16.2	44.56	22.2	62.59
4.3	11.28	10.3	27.66	16.3	44.86	22.3	62.90
4.4	11.55	10.4	27.95	16.4	45.15	22.4	63.21
4.5	11.81	10.5	28.23	16.5	45.44	22.5	63.51
4.6	12.08	10.6	28.51	16.6	45.74	22.6	63.82
4.7	12.35	10.7	28.79	16.7	46.03	22.7	64.13
4.8	12.62	10.8	29.07	16.8	46.32	12.8	64.44
4.9	12.88	10.9	29.35	16.9	46.62	22.9	64.75
5.0	13.15	11.0	29.63	17.0	46.91	23.0	65.06
5.1	13.42	11.1	29.91	17.1	47.21	23.1	65.37
5.2	13.69	11.2	30.19	17.2	47.51	23.2	65.68
5.3	13.96	11.3	30.48	17.3	47.80	23.3	65.99
5.4	14.23	11.4	30.76	17.4	48.09	23.4	66.30
5.5	14.50	11.5	31.04	17.5	48.39	23.5	66.61
5.6	14.77	11.6	31.32	17.6	48.69	23.6	66.92
5.7	15.04	11.7	31.61	17.7	48.99	23.7	67.24
5.8	15.31	11.8	31.89	17.8	49.28	23.8	67.55
5.9	15.58	11.9	32.17	17.9	49.58	23.9	67.86

Reprinted from Vogel et al., 1946, p. 164.

Table 5.3. Average extract yield from materials

	Extract yield/pound (%)
Malt (six-row)	77–80
Malt (two-row)	79–82
Wheat	97.5
Rice	80–84
Sugar	100
Corn sugar	92.5

Source: Scott Helstad and Joe Thorner, Cargill.

terial is milled using two mill settings (fine grind and coarse grind). Ground materials are put through a standardized mashing procedure. The wort collected after a specified time and temperature mashing regime is used to determine the maximum amount of extract obtained from the material and grind.

One batch of material is ground very fine, so all extractable material is accounted for. This can be considered the theoretical yield. It is the maximum extract the brewer could hope to obtain. However, a fine grind is not achievable in most breweries due to practical quality and operating considerations.

The other batch of material is coarsely ground, to more closely simulate brewery conditions. The coarsely ground material gives a more practical yield—an extract that is achievable in the actual brewery.

Laboratory yield from either fine- or coarse-ground materials may well be higher than the yield obtained in the brewhouse. The malt extract figures supplied by the maltster are based on the same type of laboratory tests described above.

b. Brewing yield is the amount of extract that the brewer can obtain, in practice, from the starting ingredients. This figure can be expressed as follows: amount of extract recovered in the wort divided by the amount of extract available in the grist times 100.

Example.   If 1,000 bbl of a 15°P wort is produced from 54,763 lb of malt with a material extract value of 0.79 lb of extract per pound of malt, the brewing yield is calculated as follows:

$$\text{Brewing yield, \%} = \frac{1{,}000\,\text{bbl} \times (41.06\,\text{bbl}_E/\text{bbl}; \text{Table 5.2})}{(0.79\,\text{lb}_E/\text{lb}_M) \times 54{,}763\,\text{lb}_M} \times 100 = 94.9$$

where bbl_E = barrels of extract, lb_E = pounds of extract, and lb_M = pounds of malt.

Note: Breweries using hammer mills and mash filters may find that fine-grind extract values closely approximate actual brewery yields.

9. How can the yield of each material used in a brew be determined?

The yield of each material used in a brew cannot be determined from the results of a mash containing more than one material, unless the pre-determined yield of all but one material is known.

Example.   A mash consisting of malt and rice is an extract from two materials of different yields. It is not possible to determine the amount of extract contributed by each material unless the yield of at least one of them is known. The yield of both materials combined can be calculated as follows. Assume a 12°P wort is made from 30 lb of malt and 11 lb of rice (a total of 41 lb of materials) per barrel. Using the formula described previously, the amount of extract per barrel is

$$\frac{(259+12)(12)}{100} = 32.52 \text{ lb}$$

If the yield of the malt is known to be 78%, the contribution of the rice to the brew can be calculated as follows. The amount of extract contributed by the malt would be

$$30 \text{ lb} \times 0.78 \text{ (yield)} = 23.40 \text{ lb}$$

The amount of extract contributed by the rice can then be determined by simple subtraction:

$$32.52 \text{ lb (total extract)} - 23.40 \text{ lb (extract from malt)} = 9.12 \text{ lb}$$

Different yields cannot be calculated if more than one is unknown.

10. How are the total materials required to produce a barrel of wort with a specified Plato determined?

Determining the total materials required to produce a barrel of wort with a specific Plato depends on the proportion of materials (based on the extract figure) chosen by the master brewer. For example, if 70% of the extract is to be derived from malt and 30% from rice to produce a barrel of 12°P wort, the calculation would be as follows.

a. Using the formula

$$\frac{(259 + \text{degrees Plato}) \times (\text{degrees Plato})}{100} = \text{Extract per barrel (in pounds)}$$

the total extract per barrel is

$$\frac{(259+12)\times(12)}{100}=\frac{271\times12}{100}=\frac{3252}{100}=32.52\,\text{lb}$$

b. If 70% of the total extract is to be derived from malt, the number of pounds of total extract is multiplied by 0.70 to find the amount of extract to be derived from malt:

$$32.52\,\text{lb}\times0.70=22.76\,\text{lb}$$

If the average yield from malt is 78%, the number of pounds of extract to be derived from malt is divided by 0.78 to determine the amount of malt required for a barrel of 12°P wort:

$$\frac{22.76\,\text{lb}}{0.78}=29.17\,\text{lb}$$

c. The same procedure is followed for rice. If 30% of the total extract is to be derived from rice, the number pounds of total extract is multiplied by 0.30 to find the amount of extract to be derived from rice:

$$32.52\,\text{lb}\times0.30=9.76\,\text{lb}$$

If the average yield from rice is 82% (*Table 5.3*), the number of pounds of extract to be derived from rice is divided by 0.82 to determine the amount of rice required per barrel:

$$\frac{9.76\,\text{lb}}{0.82}=11.90\,\text{lb}$$

d. Adding the two figures together, we have

	Raw material (lb)	Extract (lb)
Malt	29.00	22.76
Rice	11.90	9.76
Total	41.10	32.52

The percentage of materials to be used should *always* be calculated on the extract basis, because the material basis is unimportant for methods of determining yield, etc. If this is adhered to, the following formula can be applied to determine the quantity of each material required to produce a barrel of wort of any desired degrees Plato:

$(E \times P)/Y$ = Required amount of material desired

where

> E = Extract per barrel for specified degrees Plato
> P = Percentage of material desired (extract basis)
> Y = Yield of material per barrel (according to the material specification sheet)

11. What volume is used to calculate the brewhouse yield?

The number of barrels recorded in the kettle prior to striking-out represents an erroneous total, because it is measured at 212°F (100°C). As a result, it is necessary to reduce the recorded volume by approximately 4% to compensate for the contraction of the wort upon cooling. The Plato reading is observed in cooled wort (20°C [approximately 68°F]) and, therefore, represents the gravity of the cooled wort. It is practically impossible to record the Plato of hot wort, but theoretically, it would weigh less than cooled wort. Therefore, all calculations *must* be based on the quantity and gravity of *cooled* wort; any yield calculated based on the volume of boiling wort would be inaccurate.

12. How are the water requirements of a brew determined?

An average of 1.20 bbl of water is used for each barrel of wort struck from the kettle. Approximately 35–40% of the total water requirement for a brew is used in the mashing-in procedure. The exact amount is left to the judgment of the master brewer and depends on whether a thin or heavy mash is desired.

The following formula can be used to compute the approximate water requirements:

> Total water required = Number of barrels of wort × 1.20
>
> Total water requirements × 40% = Mash water volume
>
> Total water requirements – Mash water volume
> = Sparge water volume

13. How are the various mashing temperatures determined?

When determining mashing temperatures, we work in water units, and for all practical purposes, we can assume that 600 lb of materials contributes the same heat value as 1 bbl of water:

$$\frac{\text{(Number of pounds of material)}}{600} = \text{Water equivalent (in barrels)}$$

Example.   A cooker mash made from 5,000 lb of material and 50 bbl of water, with a temperature of 207.5°F (97.5°C), is added to a mash in the mash tun, which contains 10,000 lb of malt and 91 bbl of water at 95°F (35°C).

To determine the temperature of the combined mash, first convert the number of pounds of material to the water equivalent. For the mash in the cooker,

$$\frac{5,000}{600} \approx 8 \text{ bbl water equivalent}$$

$$8 \text{ bbl water equivalent} + 50 \text{ bbl of water} = 58 \text{ bbl}$$

For the mash in the mash tun,

$$\frac{10,000}{600} \approx 16.5 \text{ bbl water equivalent}$$

$$16.5 \text{ bbl water equivalent} + 91 \text{ bbl of water} = 107.5 \text{ bbl}$$

The following formula can be applied:

$$T = \frac{(V_c \times T_c) + (V_m + T_m)}{V_t}$$

where

> T = Temperature of combined mash
> V_c = Volume in cooker (water and material)
> V_m = Volume in mash tun (water and material)
> T_c = Temperature of cooker mash
> T_m = Temperature of mash in mash tun
> V_t = Total volume ($V_c + V_m$)

In this example, if V_c = 58 bbl, T_c = 207.5°F (or 97.5°C), V_m = 107.5 bbl, T_m = 95°F (35°C), and V_t = 165.5 bbl,

$$T = \frac{(58 \text{ bbl} \times 207.5°F) + (107.5 \text{ bbl} \times 95°F)}{165.5 \text{ bbl}} \text{ or } \frac{(58 \text{ bbl} \times 97.5°C) + (107.5 \text{ bbl} \times 35°C)}{165.5 \text{ bbl}}$$

$$= \frac{12,035 \text{ bbl °F} + 10,212.5 \text{ bbl °F}}{165.5 \text{ bbl}} \text{ or } \frac{5,655 \text{ bbl °C} + 3,762.5 \text{ bbl °C}}{165.5 \text{ bbl}}$$

$$= \frac{22{,}247.5 \text{ bbl } °F}{165.5 \text{ bbl}} \text{ or } \frac{9{,}417.5 \text{ bbl } °C}{165.5 \text{ bbl}}$$

$$= 134.4°F \text{ or } 56.9°C$$

The cooker mash temperature is usually tested with a thermometer as it flows to the mash tun to establish the actual temperature, because the temperature can vary considerably depending on how far from the cooker the mash tun is located.

14. What means are available for checking the following:

a. Grinding of malt. The grinding of malt is checked for uniformity using a set of standard screens. The ASBC (American Society of Brewing Chemists, 2004, method Malt-2B) recommends a minimum of three screens plus a pan with slot widths of $7/64$, $6/64$, and $5/64$ in. (2.78, 2.38, and 1.98 mm, respectively). A sample of grist from the mill is placed on the top (coarsest) screen, and the stack is shaken for a set time. The percent (by weight) retained on each screen is calculated. Variations from standard values are indicative of two things: the malt assortment has changed and/or the mill gap setting has changed.

b. Mellowness of malt. The simplest means of determining the mellowness of malt is by chewing it—mellow kernels are soft; glassy kernels are hard; and half-glassy kernels are in between. The kernel can also be cut in half, and the condition of the body observed with the aid of a magnifying glass or microscope.

A more precise method involves the use of a bench-top instrument called a friability meter. This instrument uses constant pressure and a rubber roller to press a preweighed amount of malt kernels against a rotating sieve drum. Well-modified (friable) malt is easily crushed and passes through the sieve during the test. The materials retained on the sieve tend to be unmodified steely ends, unmalted grains, and contaminants. The amount of malt that passes through the sieve is weighed and reported as percent friability (extent of modification) for the malt sample.

c. Water used in various brewhouse vessels. The amount of water used in each brewhouse vessel should be gauged or metered (see Chapter 1, Volume 3, for a discussion of meters).

d. Amount of materials used. Materials must always be weighed prior to use. Materials can be weighed by placing them on a scale, using load cells, or measuring through automatic dumpers.

e. Amount of wort in kettle.   Various methods can be used to determine wort volume, but they must be previously calibrated against an accurate water meter. A length of metal stick can be used to determine the amount of wort in the kettle. Older kettles had a stick gauge that floated on the surface of the wort and gave a good approximate reading of the kettle volume. Other methods include a flowmeter measurement of liquid going into or out of the kettle.

f. Evaporation rate of kettle.   The evaporation rate of a kettle is calculated for wort during a specified "condition" of the wort. Generally, the procedure consists of observing the quantity of wort in the kettle when the kettle is full, the wort is boiling, and the hops are fully submerged. When the wort has boiled for 30 min (with all the hops added) the quantity is again recorded. The difference multiplied by 2 indicates the rate of evaporation (in barrels of wort per hour).

g. Specific gravity of first, last, and kettle wort.   The specific gravity of the first, last, and kettle wort can be determined using a Plato hydrometer. After the Plato is calculated, the specific gravity can be determined by referencing *Table 5.1* or approximated using the following formula:

$$(\text{Plato} \times 0.004) + 1 = \text{specific gravity (approximate)}$$

Example (first wort)

$$(19 \times 0.004) + 1 = 1.076 \text{ (specific gravity)}$$

h. How is mash tested for starch conversion?   Mash conversion is tested by taking a small portion of the liquid above the grains and placing a few drops into the wells of a white spot plate. A few drops of iodine test solution are added to each test spot containing liquid. An incompletely converted mash containing starch will turn purplish black after the addition of iodine, indicating a positive test result. A completely converted mash will not change color after the addition of iodine.

15. What is iodine test solution?

One of the iodine test solutions, 0.02 N, recommended by the ASBC procedure (American Society of Brewing Chemists, 2004, Malt-4), is prepared as follows: dissolve 1.27 g of iodine crystals and 2.50 g of potassium iodide in distilled water and dilute to 500 mL. Make a fresh iodine solution every month. For daily use, keep portions of the solution in small brown dropper bottles in the dark.

16. What are brewhouse losses?

Losses are divided into two groups: (1) extract losses, determined by measurement of wort specific gravity; and (2) water losses (volume), determined by meters or gauges.

a. Extract losses

Loss of extract retained in spent grains (1–2%)
Retention of wort in spent hops (unless efficiently sparged)
Retention of a small amount of wort in trub pile
Loss of extract caused by adhesion to pipe lines, tanks, equipment (1%)

b. Water losses

Retention in grains
Retention in spent hops
Evaporation during boiling
Evaporation during cooling
Shrinkage of volume due to contraction
Loss of approximately 1% due to adhesion

17. What are processing losses (after brewhouse)?

Processing losses (shrinkage in volume) in a cold state are usually determined by metering the volume from one stage of the process to another. Losses, depending on plant design, efficiency, and type of operation, can be approximately:

Starting cellar	0.5–1%
Fermentation	
Closed	1–2%
Open	2–3%
Stock house	1.5–3%
Finishing	0.75–2%

The above losses do not include packaging and racking losses (0.5–2%).

18. What instruments are used to determine specific gravity?

a. Pycnometer. Specific gravity can be determined by weighing equal volumes of water and wort separately in a flask called a pycnometer. The weight of the wort divided by the weight of the water indicates how many times heavier the wort is compared with an equal volume of water.

A pycnometer is a glass flask with a capillary tube stopper. The contents of the flask, when filled with distilled water at 20°C, weigh approximately 50 g. The pycnometer is used to weigh an equal volume of wort and water, at the same or relative temperatures, to determine specific gravity.

b. Saccharometer (hydrometer). A saccharometer is a hydrometer constructed for determining the density of sugar solutions. Plato-scale saccharometers are generally used in breweries. They are calibrated against a cane sugar solution; for example, if the saccharometer sinks to the 10th calibration in a cooled wort, then 100 lb of the wort contains 10 lb of dry extract (based on a 10% sugar solution). A saccharometer is usually equipped with a correction scale to account for variances in temperature.

c. Refractometer. A refractometer is an instrument used to determine specific gravity (°Brix or °Plato) and refractive indices of solutions by comparing the degree to which a beam of light passing through the test solution is bent from its original path. The most common refractometer used on the brewhouse floor is a small, handheld model that resembles a short tube with an eyepiece on one end and a hinged sample glass/prism mechanism on the other end. Laboratory models are more sophisticated, include a light source, and often are temperature controlled. The advantage of a refractometer is that only a drop or two of wort is needed for the test. This instrument is only accurate with sugar and water solutions and, thus, has the same limitations as a hydrometer. Ethanol, even at the very low concentrations found at the start of fermentation, lowers the specific gravity of wort and beer, and as a result, the refractometer yields inaccurate results.

d. Densitometer. The densitometer is an instrument used to determine the specific gravity of liquids. A small capillary U-shaped tube is filled with the liquid to be tested. The machine determines the specific gravity of the liquid by oscillating the liquid and comparing the frequency of the oscillations to a known standard. Models include both tabletop and portable handheld versions.

19. What other instruments are available for determining specific gravity?

Specific gravity is directly or indirectly based on correct weight, which should be determined by a sensitive "scale" or balance. Two types of balances are used in the laboratory: the analytical balance, which is sensitive to 0.1 mg, and the technical balance, which is not as sensitive and is used for weighing heavier objects.

20. How does a glass thermometer operate?

The glass thermometer most of us are familiar with consists of a bulb, usually containing mercury, and a vacuum tube. When the bulb is heated, the liquid mercury expands (the glass does as well, but an insignificant amount) and rises in the tube. The amount of rise is calibrated on an attached scale and indicates the thermometer reading. Only thermometers containing colored alcohol, instead of mercury, should be used for general laboratory applications. The most common mercury thermometer in a modern laboratory has been precisely calibrated by the American Society for Testing Methods (ASTM) and is very, very expensive. A laboratory usually only has one or two of these thermometers, and they are used to calibrate less expensive thermometers. No glass thermometers, of any type, should ever be used in the brewery operations area. A metal-stem, bimetallic, dial-faced thermometer, or an equivalent type (see below), is safer for brewery operations.

21. Must all thermometers contain glass and liquid?

Various types of instruments can be used to measure temperature. Any measurable phenomena caused by heat can be used in thermometry.

a. Bimetallic elements bend with changes in temperatures. Using suitably connected levers and a scale, a thermometer can be made. This type is common in regulating thermostats.

b. Resistance thermometers (RTDs) respond to changes in the resistance of an electrical circuit caused by heating and cooling. They are frequently used in recording instruments.

c. Some thermometers are filled with gas instead of liquid.

d. Thermocouples generate an electric current when subjected to heat. By measuring this current, a thermometer can be calibrated.

22. What standard temperature scales are in use?

Standard calibrations, or thermometer scales, and their comparisons are illustrated in *Table 5.4*. The Celsius (centigrade) scale is the most commonly used worldwide. The Fahrenheit scale is still common in the United States. Although no longer used for temperature values in modern breweries, the Réaumur scale is included for historical purposes.

Table 5.4. Comparison of Celsius, Réaumur, and Fahrenheit thermometer scales

	Boiling point	Freezing point
Fahrenheit	212°	32°
Celsius (Centigrade)	100°	0°
Réaumur	80°	0°

Conversion of thermometer degrees

°C to °R: °C × 4/5 = °R   °R to °C: °R × 5/4 = °C   °F to °R: (°F − 32) × 4/9 = °R

°C to °F: (°C × 9/5) + 32 = °F   °R to °F: (°R × 9/4) + 32 = °F   °F to °C: (°F − 32) × 5/9 = °C

C	R	F	C	R	F	C	R	F
+100	+80	+212	+55	+44	+131	+10	+8	+50
98.75	79	209.75	55.75	43	128.75	8.75	7	47.75
97.50	78	207.50	52.50	42	126.50	7.50	6	45.50
96.25	77	205.25	51.25	41	124.25	6.25	5	43.25
95	76	203	50	40	122	5	4	41
93.75	75	200.75	48.75	39	119.75	3.75	3	38.75
92.50	74	198.50	47.50	38	117.50	2.50	2	36.50
91.25	73	196.25	46.25	37	115.25	+1.25	+1	34.25
90	72	194	45	36	113	0	0	32
88.75	71	191.75	43.75	35	110.75	−1.25	−1	29.75
87.50	70	189.50	42.50	34	108.50	2.50	2	27.50
86.25	69	187.25	41.25	33	106.25	3.75	3	25.25
85	68	185	40	32	104	5	4	23
83.75	67	182.75	38.75	31	101.75	6.25	5	20.75
82.50	66	180.50	37.50	30	99.50	7.50	6	18.50
81.25	65	178.25	36.25	29	97.25	8.75	7	16.25
80	64	176	35	28	95	10	8	14
78.75	63	173.75	33.75	27	92.75	11.25	9	11.75
77.50	62	171.50	32.50	26	90.50	12.50	10	9.50
76.25	61	169.25	31.25	25	88.25	13.75	11	7.25
75	60	167	30	24	86	15	12	5
73.75	59	164.75	28.75	23	83.75	16.25	13	+2.75
72.50	58	162.50	27.50	22	81.50	17.50	14	0.50
71.25	57	160.25	26.25	21	79.25	18.75	15	−1.75
70	56	158	25	20	77	20	16	4
68.75	55	155.75	23.75	19	74.75	21.25	17	6.25
67.50	54	153.50	22.50	18	72.50	22.50	18	8.50
66.25	53	151.25	21.25	17	70.25	23.75	19	10.75
65	52	149	20	16	68	25	20	13
63.75	51	146.75	18.75	15	65.75	26.25	21	15.25
62.50	50	144.50	17.50	14	63.50	27.50	22	17.50
61.25	49	142.25	16.25	13	61.25	28.75	23	19.75
60	48	140	15	12	59	30	24	22
58.75	47	137.75	13.75	11	56.75	31.25	25	24.25
57.50	46	135.50	12.50	10	54.50	32.50	26	26.50
56.25	45	133.25	11.25	9	52.25	33.75	27	28.75

Adapted from Vogel, et al., 1946.

Table 5.5. Boiling point of water at various altitudes

Altitude (ft)	Altitude (m)	Pressure (mm Hg)	Temperature (°F)	Temperature (°C)
0	0	759.85	212.0	100.00
500	152.4	746.54	211.1	99.50
1,000	304.8	733.46	210.2	99.01
1,500	457.2	720.60	209.3	98.52
2,000	609.6	707.98	208.5	98.03
2,500	762.0	695.57	207.6	97.54
3,000	914.4	683.38	206.7	97.05
3,500	1,067	671.41	205.8	96.57
4,000	1,219	659.65	204.9	96.08
4,500	1,372	648.09	204.1	95.60
5,000	1,524	636.73	203.2	95.12
5,500	1,676	625.57	202.4	94.64
6,000	1,829	614.61	201.5	94.16
6,500	1,981	603.84	200.6	93.69
7,000	2,133	593.26	199.8	93.21
7,500	2,286	582.87	198.9	92.74
8,000	2,438	572.65	198.1	92.26
8,500	2,591	562.62	197.2	91.79
9,000	2,743	552.76	196.4	91.32
9,500	2,895	543.08	195.5	90.86
10,000	3,048	533.56	194.7	90.39

Source: K. Loomis, New Mexico State University, Apache Point Observatory.

23. How can thermometers be tested?

Thermometers should be tested against an officially calibrated thermometer, such as an ASTM thermometer. Both thermometers are placed under the same conditions for comparison. A thermometer should read 212°F (100°C) in boiling water at sea level and 32°F (0°C) in a mixture of chipped ice and water. Adjustments for atmospheric pressure must be made for accuracy (*Table* 5.5).

24. How are thermometers maintained in good order?

Thermometers must be handled carefully; the slightest mishandling may cause a crack in the bulb, tube, or cord, causing errors in their accuracy. Thermometers are precision instruments and must be cared for in the same manner as any other delicate apparatus. Thermometers are available with protection, including metal guards and nonstick coatings. Once again: Never use a mercury-filled thermometer (even with protective coatings and shields) in a production area.

25. What is pH?

The pH (*pouvoir Hydrogène*) of a liquid is a measure of the concentration of hydrogen ions in the liquid. pH is measured on a scale of 0–14, with pH 7.0 defined as neutral, neither acidic nor alkaline. Liquids with pH values lower than 7.0 (0–6.9) are classified as acids (e.g., phosphoric acid). Liquids with pH values higher than 7.0 (7.1–14.0) are classified as bases (e.g., sodium hydroxide).

26. How is pH measured?

pH can be measured either using an instrument called a pH meter or with a section of pH test tape. The pH meter consists of an electrode probe that is immersed in the liquid to be tested. By measuring the potential difference between the electrode and the liquid in which it is immersed, it is possible to determine the concentration of specific ions dissociated in the liquid. pH test tapes are rolls of tape treated with chemical indicators that change color at specific pH levels. The pH of a liquid is determined by first wetting the test tape with the liquid to be tested. The resulting color of the tape is then compared to the colors supplied with the test tape and the (approximate) pH value is determined.

27. At what points in the brewing process is pH measured?

pH is measured throughout the brewing process to determine whether the liquid being measured has a normal value.

Abnormal values indicate possible process deviations. pH is measured at many places in the brewery, including incoming water, mash, cooled wort, fermented beer, etc. When pH values are lower than normal, microbial growth (other than brewing yeast) is usually responsible. pH measurement can also be part of the cleaning program to check final CIP rinses for completeness.

28. How is a microscope used in the examination of yeast?

A microscope is an optical instrument, consisting of a lens or combination of lenses, that enlarges, or magnifies, images of minute objects.

A microscope is used to examine yeast for purity, viability, and general physical condition. Yeast should be microscopically observed prior to and after fermentation.

29. How is yeast concentration determined?

One of the most common methods of determining yeast concentration uses a specially calibrated counting chamber (microscope slide) called a hemocytometer. The hemocytometer is divided into a series of calibrated grids. By counting the number of yeast cells in a known number of grids, the yeast concentration can be calculated.

Other modern laboratory methods of determining yeast concentration include the Coulter counter (particle size measurement) and the Aber meter (conductance by live cells).

30. How is yeast viability determined?

Yeast viability can be determined using a microscope and a special staining solution, methylene blue. A mixture of yeast and a dilute solution of the stain is placed on a microscope slide. Viable (alive) yeast cells react with and neutralize the stain; they appear clear under the microscope. Nonviable (dead) yeast cells cannot neutralize the stain; they accept the stain and appear blue under the microscope.

31. What is apparent extract?

When fermented wort is gauged by means of a hydrometer (Plato of fermented beer), the Plato reading is "apparent"—it is not a true reading, because of the presence of a new compound, alcohol. Since the Plato and Balling tables were originally based on sugar solutions, no allowance was made for alcohol. Alcohol is lighter in weight and causes the Plato reading to be relative, or apparent.

Saccharometer indication is another name for the Plato reading of the beer or the density of the beer. The saccharometer indication of the fermented wort indicates the "apparent" extract but actually shows less, as explained above.

32. What is real extract?

Real extract is the actual amount of extract remaining in fermented beer.

33. How is the amount of real extract in beer determined?

Real extract is determined by removing the alcohol present in the fermented beer by distillation. The specific gravity of the liquid, or "real extract," remaining is determined after the volume of alcohol lost by distillation is replaced by water. After the specific gravity of the resulting mixture is measured (by hydrometer, refractometer, or modern instrumentation), the corresponding grams of extract per 100 g of solution can be read from *Table 1.1* (American Society of Brewing Chemists, 2004, Tables Related to Determinations on Wort, Beer, and Brewing Sugars and Syrups, Table 1).

34. What are real attenuation and real degree of fermentation?

The difference between the Plato of the wort prior to fermentation and the actual amount of extract remaining in the beer (real extract) is called the real attenuation.

The real degree of fermentation (RDF) is the numerical explanation of the difference in starting extract and the final real extract of the beer. It can be calculated using the following ASBC formula (American Society of Brewing Chemists, 2004, Beer-6B) once the real extract value of the beer is determined.

$$\text{RDF }(\%) = \{[100(O - E)]/O\} \times \{1/[1 - (0.005161 \times E)]\}$$

where O = extract, original wort (°Plato) and E = real extract (%). The constant 0.005161 is a correction for the mass lost by CO_2 evolution and yeast uptake when the wort is fermented.

35. What are apparent attenuation and apparent degree of fermentation?

The difference between the Plato of the wort prior to fermentation and the Plato of the fermented wort (beer) is called apparent attenuation.

The apparent degree of fermentation (ADF) is the numerical explanation of the difference in the starting extract and final apparent extract of the beer. It can also be calculated using the following ASBC formula (American Society of Brewing Chemists, 2004, Beer-6C):

$$\text{ADF}(\%) = \frac{100 \times [(\text{Extract, original wort}) - (\text{Apparent extract})]}{(\text{Extract, original wort})}$$

36. How is the alcohol content of beer calculated for practical purposes?

The alcohol content of beer can be estimated by the following calculation: determine the amount of apparent extract (in °Plato) in the beer, subtract this figure from the Plato of the wort prior to fermentation, and multiply the result by 0.42. This calculation results in an approximate percentage of alcohol that is sufficiently accurate for practical purposes.

Example

Plato of wort	12
Apparent extract	− 4
Apparent attenuation	8
	× 0.42
	3.36% alcohol (approx.), by weight

37. What other means exist for a more accurate determination?

The most accurate method to determine the actual amount of alcohol in any solution is to distill the alcohol from the solution and establish the specific gravity of the distillate. This is precise work and must be accurately done with specialized laboratory equipment.

38. How is the caloric content of beer determined?

If the alcohol content and the real extract value of a beer are known, the caloric content can be determined using the following ASBC formula (American Society of Brewing Chemists, 2004, Beer-33):

$$\frac{\text{Calories (kcal)}}{100 \text{ g of beer}} = 6.9(A) + 4(B - C)$$

where A = alcohol (%, by weight), B = real extract (%, by weight), and C = ash content (%, by weight).

$$\frac{\text{Calories (kcal)}}{12 \text{ oz (355 mL) of beer}} = \frac{\text{Calories}}{100 \text{ g of beer}} \times \frac{355 \times \text{sp. gr.}}{100}$$

Note: Calories can be roughly estimated without knowing the precise ash content. Ash content is typically around 0.1% by weight.

39. How is the color of wort and beer measured?

Color of beer and wort can be measured with reference to the series 52 Lovibond scale of standard colors. Colored glass standards are matched against samples contained in glass cells. Several means exist for comparing colors in laboratories. Brewers commonly employ a series of bottles of colored solutions that have been standardized against a Lovibond instrument and are compared to the sample in question.

Laboratories with spectrophotometers can measure color by passing light through the sample at a wavelength of 430 nm.

40. What means of gauging the flow of beer are available?

Liquid meters, similar to water meters, are used to gauge the flow rate of beer (see Chapter 1, Volume 3).

41. How is the capacity of a tank determined?

The capacity of a tank can be calculated using the tank dimensions. The amount of liquid in a tank is determined by "gauging" the tank, which consists of establishing the height of the liquid and the corresponding parallel height on a previously calibrated scale. Metering into the tank and recording the barrelage on a gauge are used to calibrate the scale.

42. What are stationary and portable gauges?

A stationary gauge usually consists of a glass tube extending from the top to the bottom of a tank, parallel to which is a calibrated metal strip for comparing the liquid height with the corresponding level on the calibrated scale. A portable gauge consists of a glass tube connected on each end with rubber hosing fitted on an opening at both the top and the bottom of the tank. The liquid rises in the glass tube to the height in the tank and the level is compared to a calibrated gauge. The calibrated gauge can be used for all tanks having the same diameter and depth.

43. What apparatus is available for determining the purity of CO_2?

Breweries that collect the CO_2 produced during fermentation and brewers who purchase CO_2 can test the purity of the gas by bubbling it through a solution of sodium hydroxide. The alkali solution absorbs the CO_2 but not the air. The instrument usually has a calibrated tube for determining the percentage of air present. Generally, 99.9% purity is desired.

The CO_2 must also be free of any taints that may negatively affect the flavor of the beer. This can be tested by bubbling CO_2 through a sample of chilled water. The carbonated water is then tasted and smelled for the presence, or absence, of taints.

44. What means are available for determining the quantity of CO_2 in beer?

CO_2 volume testers are used for determining the volume of CO_2 in bottles or cans of beer and bulk beer (tanks or draft kegs). These testers are based on pressure generated in the beer by the CO_2 at specific temperatures.

The theory is based on Boyle's law for gases; i.e., at a constant temperature, the volume of a given quantity of a gas varies inversely to the pressure to which the gas is subjected.

The solubility of CO_2 in beer at different temperatures and pressures is charted in *Table 5.6*.

45. What means are available for determining the quantity of air in beer?

The determination of air in the beer bottle headspace is similar to determining the air in CO_2. To measure air in the bottle headspace, CO_2 is absorbed in sodium hydroxide, and all unabsorbed gas is considered "air." It is important to remember that the "air" remaining comprises a mixture of nitrogen and oxygen. The quantity of air is measured using a calibrated tube. Meters for detecting and measuring dissolved oxygen in beer in cellars and packaging are discussed in Chapter 1, Volume 3.

46. How are pressure gauges constructed? What are the important factors in maintaining them in good condition?

Pressure gauges are generally constructed with a "rotary movement" but depend on the fluctuation of a fine spring. They are delicate instruments and cannot be handled roughly because of the sensitivity of the spring. In some ways, pressure gauges can be compared to a watch. Corrosion must be minimized, or the gauge can stick or fail to function properly. Pressure gauges installed in the brewery should be adapted to sanitary fittings and separated from the product by a stainless membrane. Glycerine-filled gauges are commonly used to insulate the gauge from shocks.

47. How is bitterness in beer measured?

The bittering substances in beer (primarily iso-alpha acids) are extracted with solvents such as iso-octane. The extracted liquid is placed in a spectrophotometer and read at an absorbance of 275 nm. The absor-

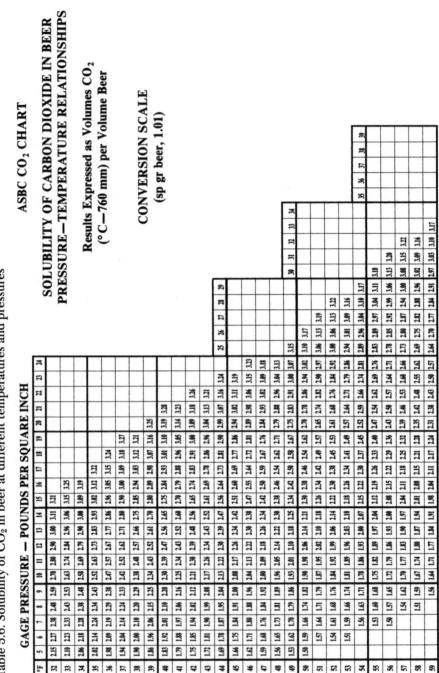

ASBC CO$_2$ CHART

**SOLUBILITY OF CARBON DIOXIDE IN BEER
PRESSURE—TEMPERATURE RELATIONSHIPS**

**Results Expressed as Volumes CO$_2$
(°C—760 mm) per Volume Beer**

**CONVERSION SCALE
(sp gr beer, 1.01)**

Table 5.6. Solubility of CO$_2$ in beer at different temperatures and pressures

GAGE PRESSURE — POUNDS PER SQUARE INCH

°F	5	6	7	8	9	10	11	12	13	14	15	16	17	18	19	20	21	22	23	24	25	26	27	28	29	30	31	32	33	34	35	36	37	38	39
32	2.15	2.27	2.38	2.48	2.59	2.70	2.80	2.90	3.00	3.11	3.21																								
33	2.10	2.23	2.33	2.43	2.53	2.63	2.74	2.84	2.96	3.06	3.15	3.25																							
34	2.06	2.18	2.28	2.38	2.48	2.58	2.69	2.79	2.90	3.00	3.09	3.19																							
35	2.02	2.14	2.24	2.34	2.43	2.52	2.63	2.73	2.83	2.93	3.02	3.12	3.22																						
36	1.98	2.09	2.19	2.29	2.38	2.47	2.57	2.67	2.77	2.86	2.96	3.05	3.15	3.24																					
37	1.94	2.04	2.14	2.24	2.33	2.42	2.52	2.62	2.71	2.80	2.90	3.00	3.09	3.18	3.27																				
38	1.90	2.00	2.10	2.20	2.29	2.38	2.48	2.57	2.66	2.75	2.84	2.94	3.03	3.12	3.21																				
39	1.86	1.96	2.06	2.15	2.25	2.34	2.43	2.52	2.61	2.70	2.80	2.89	2.98	3.07	3.16	3.25																			
40	1.83	1.92	2.01	2.10	2.20	2.30	2.39	2.47	2.56	2.65	2.75	2.84	2.93	3.01	3.10	3.19																			
41	1.79	1.88	1.97	2.06	2.16	2.25	2.34	2.43	2.52	2.60	2.70	2.79	2.88	2.96	3.05	3.14	3.23																		
42	1.75	1.85	1.94	2.02	2.12	2.21	2.30	2.39	2.48	2.56	2.65	2.74	2.83	2.91	3.00	3.09	3.18	3.26																	
43	1.72	1.81	1.90	1.99	2.08	2.17	2.26	2.34	2.43	2.52	2.61	2.69	2.78	2.86	2.96	3.04	3.13	3.21																	
44	1.69	1.78	1.87	1.95	2.04	2.13	2.22	2.30	2.39	2.47	2.56	2.64	2.73	2.81	2.90	2.99	3.07	3.16	3.24															
45	1.66	1.75	1.84	1.91	2.00	2.08	2.17	2.26	2.34	2.42	2.51	2.60	2.69	2.77	2.86	2.94	3.02	3.11	3.19															
46	1.62	1.71	1.80	1.88	1.96	2.04	2.13	2.22	2.30	2.38	2.47	2.55	2.64	2.72	2.81	2.89	2.98	3.06	3.15	3.23														
47	1.59	1.68	1.76	1.84	1.92	2.00	2.09	2.18	2.26	2.34	2.42	2.50	2.59	2.67	2.76	2.84	2.93	3.02	3.09	3.18														
48	1.56	1.65	1.73	1.81	1.89	1.96	2.05	2.14	2.22	2.30	2.38	2.46	2.54	2.62	2.71	2.79	2.88	2.96	3.04	3.13														
49	1.53	1.62	1.70	1.79	1.86	1.93	2.01	2.10	2.18	2.25	2.34	2.42	2.50	2.58	2.67	2.75	2.83	2.91	3.00	3.07	3.15													
50	1.50	1.59	1.66	1.74	1.82	1.90	1.98	2.06	2.14	2.21	2.30	2.38	2.46	2.54	2.62	2.70	2.79	2.86	2.94	3.02	3.10	3.17												
51		1.57	1.64	1.71	1.79	1.87	1.95	2.02	2.10	2.18	2.26	2.34	2.42	2.50	2.57	2.65	2.74	2.82	2.90	2.97	3.06	3.13	3.19											
52		1.54	1.61	1.68	1.76	1.84	1.92	1.99	2.06	2.14	2.22	2.30	2.38	2.46	2.53	2.61	2.68	2.76	2.84	2.92	3.00	3.06	3.13	3.22										
53		1.51	1.59	1.66	1.74	1.83	1.90	1.97	2.03	2.11	2.18	2.26	2.34	2.42	2.50	2.58	2.66	2.73	2.80	2.88	2.94	3.01	3.09	3.16	3.17									
54			1.56	1.63	1.71	1.78	1.86	1.93	2.00	2.07	2.15	2.22	2.30	2.37	2.45	2.52	2.59	2.66	2.74	2.81	2.89	2.95	3.04	3.10	3.11	3.18								
55			1.53	1.60	1.68	1.75	1.82	1.89	1.97	2.04	2.12	2.19	2.26	2.33	2.40	2.47	2.55	2.62	2.69	2.77	2.83	2.89	2.99	3.04	3.06	3.13	3.20							
56			1.50	1.57	1.65	1.72	1.79	1.86	1.93	2.00	2.08	2.15	2.22	2.29	2.36	2.43	2.50	2.57	2.64	2.71	2.78	2.85	2.92	2.99	3.00	3.08	3.15	3.22						
57				1.54	1.62	1.70	1.77	1.83	1.90	1.97	2.04	2.11	2.18	2.25	2.32	2.39	2.46	2.53	2.60	2.66	2.73	2.80	2.87	2.94	2.94	3.02	3.09	3.16	3.22					
58				1.51	1.59	1.67	1.74	1.80	1.87	1.94	2.01	2.08	2.15	2.21	2.28	2.35	2.42	2.49	2.55	2.62	2.69	2.75	2.82	2.88	2.88	2.97	3.03	3.10	3.16	3.17				
59					1.56	1.64	1.71	1.77	1.84	1.91	1.98	2.04	2.11	2.17	2.24	2.31	2.38	2.44	2.51	2.57	2.64	2.70	2.77	2.84	2.84	2.91	2.97	3.03	3.10	3.10	3.17				

																												40	41	42	43	44	45	46	47	48	49	50
60	1.54	1.62	1.69	1.75	1.82	1.88	1.95	2.01	2.08	2.14	2.21	2.27	2.34	2.40	2.47	2.53	2.60	2.66	2.73	2.79	2.86	2.92	2.99	3.05	3.11	3.18												
61		1.51	1.59	1.66	1.72	1.79	1.85	1.91	1.97	2.04	2.10	2.17	2.23	2.30	2.36	2.43	2.49	2.56	2.62	2.69	2.75	2.81	2.87	2.94	3.00	3.07	3.14												
62		1.56	1.63	1.69	1.76	1.82	1.88	1.94	2.01	2.07	2.14	2.20	2.26	2.32	2.39	2.45	2.52	2.58	2.64	2.70	2.77	2.83	2.90	2.96	3.02	3.08	3.14	3.20											
63		1.54	1.60	1.66	1.72	1.78	1.85	1.91	1.98	2.04	2.10	2.16	2.22	2.28	2.35	2.41	2.48	2.53	2.60	2.66	2.72	2.78	2.85	2.90	2.97	3.03	3.09	3.16											
64		1.51	1.58	1.64	1.70	1.75	1.82	1.88	1.94	2.00	2.06	2.12	2.18	2.24	2.30	2.36	2.43	2.49	2.55	2.61	2.68	2.74	2.80	2.86	2.92	2.98	3.04	3.10	3.16										
65			1.55	1.61	1.67	1.73	1.79	1.85	1.91	1.96	2.03	2.09	2.15	2.21	2.27	2.33	2.39	2.45	2.51	2.57	2.63	2.69	2.75	2.81	2.87	2.93	2.99	3.06	3.11										
66			1.52	1.58	1.64	1.70	1.76	1.82	1.88	1.93	1.99	2.05	2.11	2.17	2.23	2.29	2.35	2.41	2.47	2.53	2.59	2.65	2.71	2.76	2.82	2.88	2.94	3.00	3.06	3.12									
67			1.50	1.55	1.61	1.67	1.73	1.79	1.85	1.90	1.96	2.02	2.08	2.14	2.20	2.25	2.31	2.37	2.43	2.49	2.55	2.61	2.67	2.72	2.78	2.84	2.90	2.96	3.02	3.07									
68				1.53	1.59	1.64	1.70	1.76	1.83	1.87	1.93	1.98	2.04	2.10	2.16	2.22	2.28	2.33	2.39	2.44	2.50	2.56	2.62	2.68	2.73	2.79	2.85	2.90	2.96	3.02	3.08	3.15							
69				1.51	1.57	1.62	1.68	1.73	1.79	1.84	1.90	1.96	2.01	2.07	2.13	2.18	2.24	2.30	2.36	2.41	2.47	2.52	2.58	2.64	2.69	2.74	2.80	2.86	2.92	2.97	3.03	3.09	3.15						
70					1.54	1.59	1.65	1.70	1.76	1.81	1.87	1.92	1.98	2.04	2.09	2.15	2.21	2.26	2.31	2.36	2.42	2.48	2.54	2.59	2.65	2.70	2.76	2.81	2.86	2.92	2.98	3.03	3.09	3.14				
71					1.52	1.57	1.63	1.68	1.73	1.78	1.84	1.89	1.95	2.01	2.06	2.11	2.17	2.22	2.28	2.33	2.39	2.44	2.50	2.55	2.61	2.66	2.72	2.77	2.81	2.87	2.93	2.99	3.04	3.09				
72						1.54	1.60	1.65	1.71	1.76	1.81	1.86	1.92	1.97	2.03	2.08	2.14	2.19	2.24	2.29	2.35	2.40	2.46	2.51	2.57	2.62	2.67	2.72	2.77	2.82	2.89	2.94	3.00	3.05				
73						1.52	1.57	1.62	1.68	1.73	1.78	1.83	1.89	1.94	2.00	2.05	2.11	2.16	2.21	2.26	2.31	2.37	2.42	2.47	2.53	2.58	2.63	2.68	2.73	2.78	2.83	2.90	2.96	3.00				
74						1.50	1.55	1.60	1.65	1.70	1.76	1.81	1.86	1.91	1.97	2.02	2.07	2.12	2.18	2.23	2.28	2.33	2.38	2.43	2.48	2.53	2.58	2.64	2.69	2.74	2.79	2.85	2.90	2.95	3.06	3.11	3.16
75							1.52	1.57	1.63	1.68	1.73	1.78	1.83	1.88	1.94	1.99	2.04	2.09	2.14	2.19	2.24	2.29	2.34	2.40	2.45	2.50	2.55	2.60	2.65	2.70	2.75	2.80	2.81	2.86	3.01	3.06	3.11
76							1.50	1.55	1.60	1.65	1.70	1.75	1.80	1.85	1.90	1.95	2.00	2.06	2.11	2.16	2.21	2.26	2.31	2.36	2.41	2.46	2.51	2.56	2.61	2.66	2.71	2.76	2.81	2.77	2.97	3.01	3.07
77								1.53	1.57	1.62	1.67	1.72	1.77	1.82	1.87	1.92	1.97	2.02	2.07	2.12	2.17	2.22	2.27	2.32	2.37	2.42	2.47	2.52	2.57	2.62	2.67	2.72	2.77	2.72	2.93	2.97	3.02
78								1.50	1.55	1.60	1.65	1.70	1.75	1.79	1.84	1.89	1.94	1.99	2.04	2.09	2.14	2.18	2.23	2.28	2.34	2.38	2.43	2.48	2.53	2.58	2.63	2.68	2.73	2.68	2.88	2.93	2.98
79									1.53	1.58	1.63	1.67	1.72	1.77	1.82	1.87	1.92	1.96	2.01	2.06	2.11	2.16	2.21	2.25	2.30	2.35	2.40	2.45	2.50	2.54	2.59	2.64	2.69	2.64	2.83	2.88	2.93
80									1.50	1.55	1.60	1.65	1.70	1.74	1.79	1.84	1.89	1.93	1.98	2.03	2.08	2.13	2.18	2.22	2.27	2.32	2.37	2.41	2.46	2.51	2.55	2.60	2.65	2.60	2.79	2.84	2.89
81										1.53	1.58	1.63	1.68	1.72	1.77	1.82	1.86	1.91	1.96	2.00	2.05	2.10	2.15	2.20	2.25	2.29	2.34	2.38	2.43	2.48	2.52	2.57	2.62	2.57	2.76	2.81	2.85
82										1.52	1.56	1.60	1.65	1.70	1.75	1.79	1.84	1.89	1.94	1.98	2.03	2.07	2.12	2.16	2.21	2.26	2.31	2.36	2.40	2.44	2.49	2.54	2.59	2.54	2.72	2.77	2.82
83										1.50	1.54	1.58	1.63	1.68	1.73	1.77	1.82	1.87	1.91	1.95	2.00	2.04	2.09	2.13	2.18	2.22	2.27	2.32	2.37	2.41	2.46	2.50	2.55	2.50	2.69	2.73	2.78
84											1.52	1.56	1.61	1.65	1.70	1.74	1.79	1.83	1.88	1.92	1.97	2.02	2.06	2.11	2.15	2.19	2.23	2.28	2.33	2.37	2.42	2.46	2.51	2.46	2.65	2.69	2.74
85											1.50	1.54	1.58	1.62	1.67	1.71	1.76	1.80	1.85	1.89	1.94	1.98	2.03	2.07	2.12	2.16	2.21	2.25	2.29	2.33	2.38	2.43	2.47	2.43	2.61	2.65	2.69
86												1.52	1.56	1.60	1.65	1.69	1.74	1.78	1.83	1.87	1.91	1.95	2.00	2.04	2.09	2.13	2.18	2.22	2.26	2.30	2.35	2.39	2.43	2.39	2.57	2.61	2.65

Reprinted, by permission, from American Society of Brewing Chemists, 2004, *Methods of Analysis*, 9th ed., ASBC, St. Paul, Minn.

bance value, multiplied by 50, yields the number of bittering units in the sample.

48. What means are available for assessing beer foam quality?

There are many methods for testing beer foam quality. The simplest test simulates the conditions experienced by the consumer and assumes the beer is poured into a glass. The beer is poured into a clean standard beer glass using a standard pouring technique. The height of the foam is measured and carefully observed. The amount of foam generated is a direct function of the amount of CO_2 present in the beer. The time it takes for the foam to collapse is indicative of foam stability.

Numerous, more complicated laboratory methods and machines exist for foam evaluation. Since so many tests exist, no one method is the standard indicator of foam quality.

49. What is microbiological testing?

Microbiological testing refers to several methods used to evaluate samples of water, wort, and beer for the presence, or absence, of microbiological contaminants, such as bacteria and wild yeast. Over the years brewers have often used a very simple, straightforward, and inexpensive method of testing worts and beers for contamination. Samples of liquids to be tested are aseptically collected in sanitized bottles, crowned, and placed on a shelf at room temperature. The sample bottles are examined daily for the presence of turbidity or gas, which indicates microbiological contamination.

More sophisticated methods that use microbiological growth media are available for breweries with laboratories.

50. Is there a rapid method to detect biological residues on cleaned equipment?

All living organisms contain ATP. When ATP is brought into contact with the firefly reagent combination of luciferin/luciferase, a reaction takes place, resulting in the production of light that can be measured. The higher the contamination level, the higher the amount of light produced. One can determine immediately if a piece of equipment has not been cleaned thoroughly. The test allows for the instant detection of microorganisms and biological residues. Instruments that measure bioluminescence are available as tabletop or portable units.

51. What kinds of microorganisms will spoil beer?

In general, beer provides a very hostile environment to most micro-organisms. Only microorganisms that have adapted to the harsh conditions found in beer (a moderate level of alcohol, low pH, very low amount of oxygen, and dissolved CO_2 gas) will survive in the brewery.

However, there are two types of general beer spoilers that have adapted to the brewery environment: beer spoilage bacteria and a host of wild yeasts.

52. What are beer spoilage bacteria?

Beer spoilage bacteria are bacteria that can grow in the brewery and ultimately spoil beer by causing off-flavors or turbidity.

53. How many kinds of beer spoilage bacteria exist?

There are seven main kinds of beer spoilage bacteria: *Lactobacillus, Pediococcus, Enterobacteria, Acetobacter, Zymomonas, Pectinatus,* and *Megasphaera.*

Lactobacillus and *Pediococcus* species are the most common beer spoilers in breweries. The presence of CO_2 enhances their growth. *Enterobacteria* are associated with unsanitary conditions and grow primarily in worts. *Acetobacter* are most often associated with cask beers when the headspace is replaced with air instead of CO_2 or nitrogen. *Zymomonas* are primarily associated with primed beers. *Pectinatus* are rare, very difficult to isolate, and strict anaerobes. They are associated with CO_2 gas supply lines. *Megasphaera* are also strict anaerobes. They are not as common as lactobacilli and pediococci.

54. How can I identify beer spoilage bacteria?

Bacteria are classified as Gram-positive (staining purple with Gram's stain) and Gram-negative (staining red with Gram's stain). After staining, the morphology and Gram reaction are determined by microscopy. Test kits are commercially available, such as the Biolog system and API Rapid CH kits, which consist of a series of sugars that can also aid in the identification of bacteria.

55. What do bacteria look like under the microscope?
How do they spoil beer?

a. *Lactobacillus* species (lactobacilli) are short to long sticks, called rods. They are Gram-positive. These bacteria produce lactic acid and, thus, can cause souring of beer. They may also produce diacetyl and may produce turbidity if the population is very large.

b. *Pediococcus* species (pediococci) are small round balls, called cocci. They are Gram-positive. These bacteria often form groups of four cocci (tetrads), a very definitive identification marker under the microscope. They spoil beer by producing diacetyl.

c. *Enterobacteria* species are short rods. They are Gram-negative. These bacteria spoil wort by producing large quantities of dimethyl sulfide and sometimes diacetyl.

d. *Acetobacter* species are short rods. They are Gram-negative. They spoil beer by turning ethanol into acetic acid (vinegar).

e. *Zymomonas* species are short, fat rods. They are Gram-negative. The rods sometimes clump together and form a mass that resembles a rosette. They spoil beer by producing acetaldehyde and hydrogen sulfide.

f. *Pectinatus* species are slightly curved rods. They are Gram-negative. They spoil beer by producing propionic acid, hydrogen sulfide, and acetic acid.

g. *Megasphaera* species are cocci. They are Gram-negative. They spoil beer by producing butyric and caproic acids.

56. What culture media can be used to find these bacteria?

These bacteria can be grown on culture media designed specifically for the brewing industry. Some common media include universal beer agar (UBA); Wallerstein's laboratory nutrient agar (WLN); Wallerstein's laboratory differential agar (WLD), which includes cycloheximide to inhibit the growth of brewing yeasts; and Hsu's *Lactobacillus–Pediococcus* (HLP) medium.

57. What incubation conditions should be used?

The laboratory incubator should be set between 77°F (25°C) and 86°F (30°C). *Lactobacillus* bacteria grow best at the higher temperature, and *Pediococcus* bacteria grow better at the lower temperature. Often a compromise temperature of 82°F (28°C) is used.

The incubator should be flushed with CO_2 gas to provide optimal growth conditions for brewing microorganisms. If this is not possible, the laboratory can use a tight-sealing glass vessel and purchase one of the CO_2-generating gas packs available from scientific supply houses. Simply burning a candle in the vessel will remove the oxygen from the air, but will not provide the CO_2-rich environment needed.

58. What is wild yeast?

Any yeast in a brewery, other than the selected brewing yeast, is defined as wild yeast. Common wild yeasts include *Saccharomyces diastaticus* and species of *Hansenula*, *Pichia*, *Debaryomyces*, *Kluyveromyces*, and *Schizosaccharomyces*.

59. Why are wild yeasts considered beer spoilers? What do they look like?

Wild yeasts are classified as beer spoilers because they are capable of producing various off-flavors (fruity, banana, higher alcohols) in beer. They may also cause turbidity.

Wild yeasts often look very different from brewing yeast. However, some wild yeasts are indistinguishable from brewing yeast when viewed using a microscope.

60. What culture media can be used to detect wild yeast?

Two general types of media are used. Brewery culture yeast will not grow on either of these media.

a. Lysine media.   These media contain lysine as the only nitrogen source. Brewing yeast will not grow on lysine media, but wild yeasts, such as species of *Hansenula*, *Pichia*, *Debaryomyces*, *Kluyveromyces*, and *Schizosaccharomyces*, can grow on them.

b. Crystal violet media.   Crystal violet inhibits the growth of brewing yeast. Brewing yeast will not grow on these media, but wild *Saccharomyces* yeasts, such as *S. diastaticus,* will grow on them.

61. What incubation conditions should be used?

The same conditions used to incubate bacteria can be used to incubate plates for the detection of wild yeast. Alternatively, incubation conditions may be aerobic. The laboratory incubator should be set between 77°F (25°C) and 86°F (30°C). Often a compromise temperature of 82°F (28°C) is used.

62. What means are available to train sensory panelists?

Panelists are trained to identify a wide variety of flavor notes by tasting samples "spiked" with the flavor note under evaluation. Panelists vary in their ability to sense flavor notes. Some are very sensitive to a particular flavor, while others may be "blind" (unable to detect) to the same flavor.

Spiked samples can be prepared from food-grade chemicals. They can also be prepared by dissolving a "tablet" obtained from a sensory training kit into a set volume of beer.

63. What samples should be tasted daily?

The following samples should be tasted daily: incoming brewing waters, raw materials, worts and beers in process, and finished beers. If possible, it is best to have several trained people tasting samples for the presence, or absence, of taints.

Note

This chapter has dealt with methods and instruments used in brewing control, but it must be understood that all these methods, calculations, and apparatus are insufficient for control of the production of quality beer. By far the most important factors in this respect are the senses of the master brewer pertaining to sight, smell, and taste.

The capability of the master brewer to investigate, interpret, and apply the results obtained through the means and methods available for brewing control and development constitutes the art of brewing.

REFERENCES

American Society of Brewing Chemists. 2004. *Methods of Analysis.* 9th ed. Malt-2B, Physical Tests: Assortment; Malt-4, Extract; Beer-3, Apparent Extract; Beer-5, Real Extract; Beer-6A, Calculated Values, Extract of Original Wort; Beer-6B, Calculated Values, Real Degree of Fermentation; Beer-6C, Calculated Values, Apparent Degree of Fermentation; Beer-13, Dissolved Carbon Dioxide; Beer-33, Caloric Content (Calculated); Tables Related to Determinations on Wort, Beer, and Brewing Sugars and Syrups. ASBC, St. Paul, Minn.

Vogel, Edward H., Jr., Schwaiger, Frank H., Leonhardt, Henry G., and Merten, J. Adolf. 1946. *The Practical Brewer: A Manual for the Brewing Industry.* Master Brewers Association of America, St. Louis, Mo.

Index